THE IRON ROAD

THE IRON ROAD

AN ILLUSTRATED
HISTORY OF THE RAILROAD

CHRISTIAN WOLMAR

LONDON, NEW YORK, MUNICH, MELBOURNE, DELHI

Editor	Hugo Wilkinson
Senior Art Editor	Phil Gamble
Editorial team	Andy Szudek
	Georgina Palffy
	Alison Sturgeon
	Richard Gilbert
	Margaret Parrish
	Rebecca Warren
Designers	Paul Drislane
	Katie Cavanagh
	Stephen Bere
Senior Editor	Angela Wilkes
Picture Research	Luped Media Research
Jacket Designer	Mark Cavanagh
Jacket Editor	Maud Whatley
Jacket Design Manager	Sophia M.T.T.
Producer, Preproduction	Rebecca Fallowfield
Production Controller	Mandy Inness
Managing Editor	Stephanie Farrow
Senior Managing Art Editor	Lee Griffiths
Publisher	Andrew Macintyre
Art Director	Phil Ormerod
Associate Publishing Director	Liz Wheeler
Publishing Director	Jonathan Metcalf

First American edition, 2014
Published in the United States by
DK Publishing
4th floor, 345 Hudson Street
New York, New York 10014

14 15 16 17 18 10 9 8 7 6 5 4 3 2 1
001–256501–April/14

Copyright © 2014 Dorling Kindersley Limited. All rights reserved

Published in Great Britain by Dorling Kindersley Limited.

A catalog record for this book is available from the Library of Congress.

ISBN 978-1-4654-1953-8

DK books are available at special discounts when purchased in bulk for sales promotions, premiums, fund-raising, or educational use. For details, contact DK Publishing Special Markets, 345 Hudson Street, New York, New York 10014 or SpecialSales@dk.com.

Printed and bound by
South China Printing Company, China.

Discover more at
www.dk.com

Contents

178   RAILROADS COME OF AGE

258   WAR AND UNCERTAINTY

328   THE IRON ROAD TODAY

Introduction

Of all the great inventions of the Industrial Revolution, the railroads had the most impact. In a world before rail, travel over any distance was a major undertaking. Regions of even small countries, such as France or England, could be up to seven days' journey from the capital, while large countries like the United States, China, or Russia could take months to cross. Until the early 19th century, most people lived their whole lives within the confines of the town or rural area in which they were born, and no one had ever gone faster than a horse could gallop. Travel was simply too difficult and too expensive for the vast majority of people, which in turn limited the spread of ideas and technology.

Lack of mobility was a major barrier to economic and social development. In the absence of rapid transportation, people could starve within only a few hundred miles of plentiful food supplies. The slow transit of goods by horse and cart, or along rivers and canals, meant that perishables had to be consumed very rapidly. Sending a letter across the country took days, and newspapers were a misnomer, since they were effectively full of old information. It took months for people to learn the fate of loved ones at war, and news of major events, from even just a short distance away, filtered through slowly.

Limitations on travel also translated into social restriction—people had little choice of spouse, since opportunities to meet potential marriage partners from even neighboring towns were rare. Concepts of time, too, were different before the railroads. Daily life was regulated by the sun, and towns just a few miles east or west of each other could operate on different schedules.

Then the railroads arrived and changed everything. One of their first major impacts was to force countries to standardize their measurements of time, both nationally and internationally, since railroad timetables would otherwise be too confusing. Greenwich Mean Time, the standard by which the world sets its clocks, was created partly as a result of the railroads, and the US was divided into four time zones for the same reason. The Trans-Siberian, the longest railroad in the world, still operates according to Moscow time, even though the line crosses seven time zones on its way to Vladivostok. Punctuality and time-keeping thus became vital, not just with regard to the railroads, but in all spheres of life. The railroads created the structured day, which, prior to their arrival, had been ten hours long, rather than today's norm of eight. In other words, they created the "nine-to-five" routine.

The railroads overthrew all established concepts of distance and time, and social upheaval followed. The last vestiges of feudalism were swept away, since people were no longer tied to the land—indeed, they could now work far from home. People worked standard hours and expected to do so for a wage. Thus, the growth of capitalism went hand in hand with the expansion of the iron road. As people no longer had to find employment near their homes, towns and cities could become far larger than would previously have been possible. Suburban sprawl, often thought of as a product of the automobile age, is, in fact, the result of the development of commuter lines.

Access to long-distance travel, in relative comfort at fairly low cost, changed people's horizons and opened up their imaginations. What had previously been impossible became routine—such as going to the seaside or visiting an exhibition. On a social level, the scope for potential marriage partners suddenly broadened, being no longer confined to the immediate vicinity. The exchange of ideas took off as national conferences could now be convened, and the inventions of the Industrial Revolution spread—first across Britain, then across the world. Professional sports became feasible as teams and their supporters could travel long distances to play other teams, and league size was limited only by how far a team could travel in a day.

Warfare, too, was revolutionized by the railroads. Armies had traditionally sustained themselves by foraging and pillaging—an unreliable practice that made it impossible to keep troops in the same place for long, since supplies, especially for the animals, inevitably ran out. These logistical restrictions meant that battles took place over days rather than weeks or months. With the rise of railroads, armies no longer needed to be constantly on the move, since they could be supplied with food and munitions from the nearest railhead. They were also invaluable in transporting troops quickly to quell domestic riots or launch wars against neighbors.

Nation states became more cohesive as country-wide railroad systems developed. The railroads, often state-owned, were the glue that bound a country together, linking disparate regions and enabling governments to expand their influence in remote, previously lawless areas. The railroads also stimulated large movements of people: Siberia and the American West were both populated after major lines were built. Settlements everywhere congregated around the tracks; indeed, in the US, several towns that were bypassed by the iron road simply moved to be closer to it. Stations became hubs, attracting development and commerce.

As the railroads expanded they brought change in their wake. Railroad companies were often the largest organizations in their respective countries and, due to their size, required new

STEAM AND SPEED
Built in 1938, the *Mallard* was the fastest steam engine of all time, and came to symbolize the technological triumphs of the golden age of the railroads.

types of business management and even accounting methods. The very engines of capitalism—bank loans, stock markets, information on investment—suddenly became possible. Railroad companies needed banks to fund their expansion and, in turn, banks found railroad companies to be their best clients, since they were the biggest and most ambitious. It was no coincidence that banks and railroad companies were the driving force of mid- and late-19th century capitalism.

Moreover, since the railroad companies employed more workers than any other, it was inevitable that, as the trade union movement was born and developed, the railroads became the industry in which they flourished. Indeed, they were the site of many of the fiercest disputes between capital and labor.

The story of the railroads is not just one of trains and technology, and this richer history, set in a wider social context, is the one this book aims to tell. Despite a strong challenge from the automobile, the railroads remain a brilliant technological feat and a great way to travel—but they are, in fact, much more than that, as every page will show.

The First Tracks

LIMMAT
STEAM, 1847

Today's railroads are a combination of inventions that were made over millennia, starting with the wheel around 8000 BCE and culminating in the steam engine in the late 18th century. By the early 1800s, steam engines—which began as huge, cumbersome machines—were small enough to be put on wheels, and so the self-propelling "steam locomotive" was born. The next stage was simply to hitch the locomotive to a train— consisting either of cars carrying freight, or of coaches bearing passengers.

There were many who argued that train travel would never be popular, or that horses should provide the power, but once the world's first major railroad, the Liverpool and Manchester line, opened in England in 1830, there was no stopping the spread of the iron road. The United States quickly followed, and the invention spread throughout Europe—tentatively at first, but then quite rapidly. Lines were opened to enormous fanfare, and people flocked to the new stations—many to begin routines of journeying to work by rail, many just to enjoy the new technology.

There were some initial setbacks, however. Every aspect of the railroads, from laying tracks and signaling to training staff and building stations, had to be learned from scratch. This was the birth of a completely new industry and teething problems were inevitable. There were accidents and fires, and investors fell prey to fraudsters and confidence tricksters. Indeed, locomotives were prone to blow up or break down, and a prominent politician, William Huskisson, was killed at the opening of the first railroad when he failed to respond quickly enough to warnings of a passing train. Nevertheless, all these difficulties were overcome, and within a couple of decades trains were traveling at twice the speed of a galloping horse and covering huge distances. The railroad age had begun.

From Wagonways
to Railroads

THE WORLD'S FIRST RAILROAD, the Liverpool and Manchester line, opened in England in 1830, at the end of the Industrial Revolution. It was the culmination of decades of experimentation with tracks, wagons, and engines, and proved beyond doubt that the railroad was the future of land transportation. But it was also an ancient technology. The wheel had been invented more than 7,000 years earlier, and had soon been given a track. By the time of the Ancient Greeks (600 BCE–600 CE), the wheels of carts and coaches were running in specially dug out channels that prevented them from sliding off the road in wet weather, and similar tracks have been found in the ruins of Pompeii and Sicily. Myth has it that Oedipus slew his father when the two came across each other on one such road and argued over who had the right of way.

The earliest image of wooden tracks being used for transportation dates back to 1350 and can be seen in a church in Freiburg im Breisgau, Germany. Within a couple of centuries, numerous wagonways (or paths made of such tracks) had been built in Germany and Britain to haul heavily loaded wagons out of mines. The first of these appeared in Saxony, which had become a major tin and silver mining region by the 14th century. Activity in the Saxon mines peaked in the 16th century, thanks to the development of the *Leitnagel Hund*—a four-wheeled mining truck that was far more efficient than its predecessors. It had an iron bar that projected from its underside into a groove between a pair of wooden tracks, to keep it from veering off course. Operating the truck demanded great skill, and there were inevitably accidents, but it soon revolutionized the German mining industry, making it possible to transport much larger quantities of ore to the surface for smelting. At first, this system was entirely dependent on manpower, but soon horses replaced men, enabling even heavier loads to be moved.

The next development was the introduction of rails for the trucks to travel on. The earliest of these, found in Germany and known as *Karrenbahnen*, were made of wood, and by the early 18th century, in the coal region of the Ruhr, they had a lip—an L-shaped flange that kept the wagons on the tracks. On some wagonways, the flange was fitted to the wheels of the trucks rather than the tracks, an arrangement that later became standard on the railroad.

By the time the flange had been introduced, Britain had also developed a system of wagonways. It was based on the German system, but soon became more extensive than its precursor. Britain was the cradle of the Industrial Revolution, and its wagonways connected an ever-expanding network of mines to an increasing number of factories, and to waterways that enabled coal and minerals to be shipped even further afield. This transportation system had a huge economic impact on Britain, and both the industrial and domestic consumption of coal increased tenfold between 1700 and the early 1800s. The network that emerged in the northeast of England in the 17th century was so busy with traffic that it became known as the Newcastle Roads. By 1660, there were nine wagonways on Tyneside alone, and several others in the Midlands to the south. In 1726, a

THE FIRST TRACKS
Misrah Ghar il-Kbir in Malta is the site of limestone cart tracks dating to around 2000 BCE. In places the ruts cross each other and form junctions, giving the appearance of tracks at a railroad switching yard.

group of coal mine owners called the Grand Allies went a step further by linking their collieries to a shared wagonway. They even created a "main line," the Tanfield Wagonway, much of which had two tracks, permitting a continuous flow of inbound and outbound vehicles. The route linked several pits with the River Tyne, crossing the Causey Arch—a bridge with a 105-ft (32-m) span over a rocky ravine—en route. Costing £12,000 (around £1.5 million or $2.4 million in today's money), the arch was built by stonemason Ralph Wood and still stands today. At the time it was the longest single-span bridge in Britain, and it is perhaps the oldest railroad bridge in the world. It accommodated two tracks—the "main way" to take coal to the river, and the "bye way" for returning empty wagons—and at its peak, over 900 horse-drawn wagons crossed the arch each day. There were several such wagonways in Britain and the rest of Europe, but cooperative ventures were rare—many pit owners deliberately built wagonways that prevented rivals from reaching the waterways.

It was not until the late eighteenth century that iron rails were first used, notably by mining engineer Friedrichs zu Klausthal near Hanover, Germany. Soon afterward, iron rails were laid to move trucks around the ironworks at the key industrial site of Coalbrookdale in the Midlands, England. The initial idea was to cover existing wooden rails with an iron cap so that they would last longer (they had previously been replaced every year), but advances in smelting technology soon made it possible to construct the entire rails from iron. It was around this time that the words "railroad" and "railway" were adopted for the wagonways of the Midlands, which now carried lime, ore, and pig iron as well as coal. Throughout this period, railroad cars also became larger and were able to carry loads of more than 2½ tons (2.5 tonnes), and the gauge of the tracks (the distance between them) was largely

TOP SPEEDS THEN AND NOW

Nicholas Cugnot's locomotive, 18th century

2½ MPH
(4KPH)

Train à Grande Vitesse (TGV), 21st century

357 MPH
(574KPH)

standardized at 5ft (1.5m). This width best suited the horses that hauled the cars (any wider and they were too heavy to pull), and it was close to the gauge that eventually prevailed across much of the world—4ft 8½in (1,435mm), or today's "standard" gauge.

These iron railroads flourished for about 40 years. At their peak, thousands of miles of iron tracks stretched across Britain, as opposed to the mere hundreds of miles of wooden equivalents that preceded them, and they reached far beyond the coalfields, linking mines and quarries with ports, rivers, and canals. Their main purpose was to transport minerals to the nearest waterway, but a few carried passengers on a casual basis—usually workers hopping on for a ride to or from a mine or a quarry. Some lines, such as the Swansea and Mumbles line in South Wales, provided passenger cars for people, but the main business was freight.

A few canny mine owners devised more sophisticated railroad systems, involving cables and the use of gravity. Ralph Allen, the owner of a quarry above Bath Spa, designed a wagonway on which the loaded cars descending the hill into the city pulled the unloaded ones back up the incline behind them. Such "gravity railroads" became common in the 18th century. The simplest ones used gravity to roll the cars downhill, and horses to pull the empty ones back up. An entirely frivolous example was the *Roulette*, built for Louis XIV of France in the gardens of Marly, near Paris. The Sun King liked to entertain his guests by giving them a ride on the *Roulette*. Although technically a gravity railroad, this was really more like a rollercoaster built into a hill. The carved, gilded car thundered down an 820-ft (250-m) wooden track into a valley and then shot up the other side. The passengers boarded from a small, classical building that could lay claim to being the world's first train station, then three bewigged valets pushed the car to the top of the incline, from where gravity took over, giving the aristocrats a novel thrill.

For all their variety, the early railroads still required horses or humans to pull them along. What they needed was an engine to drive them, and just such a device was being developed. This was the steam engine, which began life as a stationary machine that generated power to drive water pumps, but was soon adapted to provide rotary power to drive wheels. It was a small leap to connect the engine to the wheels it was driving, thus creating a self-propelling, steam-powered vehicle.

THE CAUSEY ARCH
Shown here in a watercolor by Joseph Atkinson, the Causey Arch, built 1725–26, is thought to be the oldest surviving single-arch railroad bridge in the world. The line has long since closed, but the bridge still serves as a footpath.

The idea of steam power goes back to Classical times—Archimedes recognized it, and Hero of Alexandria experimented with it—but it was only in the late 17th century that Frenchman Denis Papin harnessed it with his "steam digester," a crude pressure cooker that he adapted to make an "atmospheric engine"—essentially a cylinder containing a piston that could oscillate under the pressure of steam expanding and condensing. Applying the principles that Papin had documented, Thomas Newcomen, an ironmaster from Devon working in the early 18th century, developed the idea of producing steam engines to pump water from mines, and made 60 of them. After Newcomen died and the patents ran out, engineers copied his ideas and his engines were built in many countries, including the US, the German states, and the Austrian Empire, where one was used to power the fountains for Prinz von Schwarzenberg's palace in Vienna. However, it was Scottish engineer James Watt who first made steam power commercially viable, by making a series of technical improvements to Newcomen's designs so that they could be adapted to carry out a wide variety of tasks. He formed a partnership with English manufacturer Matthew Boulton, and soon their engines were powering looms, mills, and ships across the world.

Nicholas Cugnot, an artillery lieutenant from eastern France, made the first attempt to put a steam engine on wheels around 1672, when he designed what he hoped would be a motorized gun platform. On a test run in Paris, his engine propelled itself slowly forward, but then it veered off course and overturned, prompting the city authorities to ban it as a public danger. There were various other attempts, both in Britain and the US, to build steam-powered road vehicles, but they were so heavy they destroyed the roads—a problem that was solved by Cornishman Richard Trevithick, who first put the

FRENCH STEAM PIONEER
Denis Papin invented the "steam digester",
a type of pressure cooker that was designed
to soften bones. It was the first device to
use the power of steam, and led Papin to
create the first steam-driven engine.

THE "STEAM TROLLEY," 1672
Nicholas Cugnot's steam-powered
engine on wheels is thought to
have been the first self-propelled
vehicle in the world.

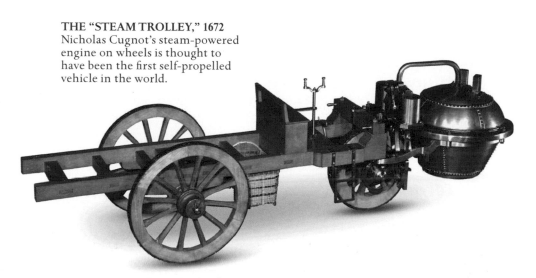

engine on rails. Trevithick had a setback in 1801, when his "road
carriage" caught fire, but three years later he produced a locomotive
that traveled at 5mph (8kph) at Pen-y-Darren, an ironworks in Wales.
Later still, he demonstrated his invention on a circular track near the
present site of Euston station, London, playfully calling it *Catch Me
Who Can*. Like Louis XIV's *Roulette*, however, it was more of a
amusement park ride than a serious commercial enterprise, and
Trevithick went off to develop stationary steam engines for the gold
and silver mines of Peru.

By the time Trevithick left England, all the technology needed
for a modern railroad was in place. Like railroad tracks, the steam
locomotive was slow to develop—indeed, it seemed for a while that
horses would power the railroads. It took time for new methods to
replace the old, and for many years the railroads were a patchwork of
iron and wooden tracks, with steam locomotives being tested while
horse-drawn lines were being built. What the railroads needed was a
genius to bring these disparate elements together. That genius was
civil engineer George Stephenson—a great synthesizer of ideas, who
was soon dubbed the "father of railroads."

The Father of Railroads

HUNDREDS OF INVENTORS, at different times, contributed to the creation of the railroads, but one individual stands out, not because he was the cleverest or the most innovative, but because he was the best at exploiting the ideas of others and turning them into workable concepts—George Stephenson. Born in modest circumstances in Wylam, near Newcastle-upon-Tyne, England, he was a barely literate, self-educated man who did not suffer fools gladly, but who deserves the title of "father of railroads", even though he made no significant inventions of his own. He played an important role in the development of two important lines: the Stockton and Darlington, completed in 1825, which was really the last and most sophisticated of the wagonways (see pp.14–21); and the Liverpool and Manchester, whose opening five years later heralded the real start of the railroad age. He continued to play a vital role in the spread of the railroads, both in Britain and abroad, until his death in 1848.

Stephenson started out as a pit boy, but he soon realized that in order to progress he needed engineering skills, and he learned these initially with the help of a local schoolmaster. Before long, he found work as an engine wright and was put in charge of all the stationary engines at Killingworth, a large coal mine in Northumberland. He quickly realised that the key to making better use of steam technology was to enable the engines to run on rails and haul loads directly, rather than using the cable system, whereby stationary engines reeled in cables attached to cars (see p.71), and he persuaded the colliery

GEORGE STEPHENSON
A self-taught engineer, Stephenson pioneered a steam locomotive that ran on rails, and built the world's first fully fledged steam railroad, the Liverpool and Manchester line, which opened in 1830.

**LIVERPOOL AND MANCHESTER
TRAIN TICKET, c.1830**
Passenger train cars on the Liverpool
and Manchester were originally divided
into first, second, and third class.
Despite the basic conditions in third,
the railroads represented a new era
of mobility for ordinary people.

owners to give him the means to build a "traveling machine"—his
name for the locomotive. The result was unveiled in 1814—a steam
engine somewhat oddly named *Blücher* after Gebhard Leberecht von
Blücher, a Prussian general who led his army to several victories
against Napoleon, the common enemy of Britain and Prussia at the
time. Incorporating features from Trevithick's engines (see pp.20–21),
Blücher was built in the colliery shop and proved to be more powerful
than any of its predecessors, as it could haul 30 tons (30.5 tonnes)
of coal up an incline at 4mph (6.5kph).

The success of *Blücher* spurred Stephenson to build another 16
locomotives over the next 10 years, most of which were commissioned
by local collieries. One went to the Kilmarnock and Troon Railway in
1817, but was withdrawn because it damaged the line's cast-iron rails.
The same fate befell a second engine that was sent to the railroad at
Scott's Pit at Llansamlet, near Swansea, in 1819. These failures
demonstrated the great difficulty in developing engines that were light
enough not to break the primitive tracks, but powerful enough to
haul a reasonable load. Showing his versatility as an engineer,
Stephenson initially solved this problem by using steam pressure to
create a "steam spring" to cushion the weight of the load. He then
simply increased the number of wheels to distribute the weight. In
1820, Stephenson was hired to build an 8-mile (13-km) railroad at
Hetton colliery. The result was a train that relied on gravity on the
downward slopes and steam locomotion on the level or uphill sections.
It was the first railroad that used no animal power whatsoever.

However, Stephenson's engines continued to have difficulties, and
he did not have the resources to solve them. In the early 1820s, he
became quite despondent, but then developments in the coal town of
Darlington lifted his spirits. A group of prosperous Quaker colliery
owners, led by Edward Pease and his son Joseph, wanted to create a

GEORGE STEPHENSON'S RAILROAD GAUGE OF

4FT 8½IN
(1,435 mm)

BECAME THE WORLD STANDARD

railroad that would connect Darlington to Stockton at the mouth of the River Tees, where coastal shipping from London docked. They wanted to reduce the price of coal by making it cheaper to transport, and to counter an alternative plan that was being discussed to build a canal. Stephenson was the obvious man to prepare such a route and build the Stockton and Darlington Railway, and was summoned to the Peases' home to discuss their plan. He was duly appointed surveyor and engineer on the project, and, since he had formed a company in 1823 with his son, Robert, to build locomotives at a works in Newcastle, he could use his own engines on the railroad. Nevertheless, when Stephenson surveyed the area to be crossed, he encountered considerable opposition from local landowners and had to map out a route that would avoid their fox-hunting grounds.

The plan was far more ambitious than any of the former wagonways. Nearly 37 miles (60km) of track had to be laid, and there were major physical obstacles too, notably the Myers Flat swamp and the Skerne River at Darlington. Stephenson eventually created a firm base in the swamp by filling it with tons of hand-hewn rock, and called on a local architect to help him design a stone bridge to cross the river. Despite its length and logistical difficulties, the line took only three years to construct, but even as it opened debate raged over what form of traction to use. Stephenson and his son produced the steam engine *Locomotion No. 1*, which, on the opening day of June 27, 1825, pulled a train of 34 cars carrying 600 passengers and a variety of goods through the countryside. However, this was not enough to convince the Peases. The truth was that Stephenson's locomotives were unreliable—they often ran out of steam and frequently needed repairs—so most of those early trains were hauled by horses; at one point, the Peases even considered turning the whole line over to horse-driven trains. Eventually, however, a much better engine designed by an engineer at Stephenson's works, Timothy Hackworth, saved the day, and it was the horses that were phased out.

The completed Stockton and Darlington Railway was recognized as a major technical advance over its predecessors, but, given the use of horses, it was still effectively a superior type of wagonway. It was also flawed. It had few sidings—which allow trains traveling in opposite directions to pass each other—so arguments and even fights between drivers were common. The owners also made the mistake of allowing anyone who was prepared to pay a fee to use their vehicles on the line, which meant that all kinds of conveyances, rickety and unstable, were used, resulting in frequent breakdowns. Nevertheless, the railroad attracted a lot of traffic, and, although it took time to become profitable, it established an important precedent—Stephenson had decided on a gauge of 4ft 8½in (1,435mm), which became standard across much of the world's railroad network.

The heavy traffic on the Stockton and Darlington line encouraged entrepreneurs across Britain to promote local railroad plans. Many of these were never built, but the most important became Stephenson's next big project—the Liverpool and Manchester line, which was conceived on a much larger scale than the Stockton and Darlington and would become the world's first modern railroad. A group of wealthy industrialists in the northwest of England, annoyed at paying local canal owners' extortionate rates to transport their goods, sought to link the two towns with a railroad. The 31-mile (50-km) line was much more ambitious than the Stockton and Darlington, and although originally conceived as a

LOCOMOTION NO. 1
This is a model of the locomotive built by Robert Stephenson and Company in 1825, which opened the Stockton and Darlington Railway, hauling passenger and freight cars.

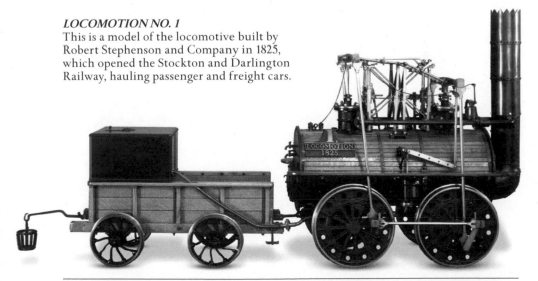

freight railroad, it would also carry passengers, as it linked two very important towns that had a combined population of 350,000. To ensure reliability, the trains would be run directly by the company, which set the pattern for nearly all the lines of the future.

Stephenson was invited by the directors to determine the route. Again, there was a major area of swampy peat bog, Chat Moss, as well as a series of streams and rivers that needed fording, requiring no fewer than 64 bridges, including a nine-arch viaduct over the River Sankey. Nothing on this scale had ever been attempted before. Stephenson was again both surveyor and engineer, and personally studied the terrain. Local landowners violently opposed the line, and there was a false start when the initial Parliamentary Bill (which needed to be passed for the line to be built) was thrown out at the behest of the rival canal owners—partly a result of an embarrassing performance by Stephenson, who proved to be somewhat

inarticulate in the intimidating atmosphere of Parliament. For a time, Stephenson was replaced, but he was soon reinstated, and work began in 1827. Stephenson personally oversaw construction all along the line, often riding long distances on horseback to check on progress. His solution to the Chat Moss problem was to float the railroad embankment on a bed of brushwood and heather. He also excavated a 2-mile (3.2-km) cutting through Olive Mount at the entrance to Liverpool, and built the Sankey Viaduct with sandstone blasted out of the cutting.

The form of traction to be used remained an issue throughout the line's construction. The directors of the railroad favored locomotives, but were unsure whether the existing engines were up to the task, so they launched a competition, the Rainhill Trials, to decide who should design the engines for the railroad. Five entrants took part in the trials, which were held on a completed section of track on

SANKEY VIADUCT
The viaduct over the River Sankey was one of the trickiest engineering challenges that George Stephenson faced in constructing the Liverpool and Manchester Railway.

October 6, 1829, in front of 15,000 spectators. The technical requirements were strict, particularly regarding weight, which was fixed at a maximum of 6 tons (6 tonnes), and the engines had to complete twenty 2-mile (4-km) trips at an average speed exceeding 10mph (16kph).

One of the entrants, *Cycloped*, turned out to be a prank (the engine was in fact a horse on a treadmill), so *Rocket*, the entry of George Stephenson's son, Robert, faced only three rivals: John Braithwaite and John Ericsson's *Novelty*, Timothy Hackworth's *Sans Pareil*, and Timothy Burstall's *Perseverance*. In the event, the trial was easily won. *Perseverance* never managed more than 6mph (10kph), and the other two failed to finish the course. Meanwhile, *Rocket* thundered up and down the track, ensuring that the Stephensons won the £500 prize to help develop their engines.

Since the line carried freight in both directions—the raw materials from the port at Liverpool, and the goods being brought back—as well as passengers, it was double-tracked from the outset, which greatly increased capacity. The opening day on September 15, 1830 was an epoch-making event, attracting people from around the

CYCLOPED
In the 1829 Rainhill Trials to find the best locomotive, *Cycloped* was proposed as an alternative to the steam engine. Powered by four horses on a treadmill, it was disqualified from the trials, which were won by Stephenson's *Rocket*.

"George Stephenson told me as a young man that railways will supersede almost all other methods of conveyance"

JOHN DIXON, QUOTED IN *LIFE OF GEORGE STEPHENSON*, 1875

world, several of whom would return home to inspire the building of railroads in their own country. The celebrations, however, were marred by tragedy—an accident that resulted in the death of William Huskisson, a prominent politician. Huskisson crossed the tracks to greet the Prime Minister, the Duke of Wellington, when the ceremonial trains stopped at Parkside, halfway down the line. Panic set in when another train, *Rocket*, approached. Huskisson failed to climb onto the Duke's car in time, and fell under the oncoming train. His leg was shattered and, although Stephenson took him to Manchester for help—reaching an amazing speed of 35mph (56kph) en route—Huskisson died that evening.

Despite this tragedy, the completion of the Liverpool and Manchester marked the peak of Stephenson's career, but it was by no means its end. He went on to build numerous railroads and his son, Robert, who concentrated mainly on improving the locomotives produced by Robert Stephenson and Company, constructed a far longer railroad, the 112-mile (180-km) London and Birmingham line, which is now part of the West Coast Main Line. The locomotive works thrived and eventually produced more than 3,000 engines before being absorbed by a larger company in 1937. George Stephenson also advised early American rail promoters (see pp.34–35), and assisted in the construction of lines in Belgium (see pp.43–44) and Spain. The Stephensons certainly left their mark on the railroads. When the line celebrated its 150th anniversary in 1980, the then-chairman of British Rail observed: "The world is a branch line of the Liverpool and Manchester."

Powering the Engine

Steam was recognized as a potential energy source as early as the 1st century CE, when steam-powered devices appeared in the writings of Hero of Alexandria. But it was not until the late 18th century that steam power was put to practical use in the form of stationary engines. The principles behind these engines were refined simultaneously by Owen Evans in the US and Richard Trevithick in England. Trevithick developed the idea of using high-pressure steam, which allowed the engine to be small enough to be mounted on wheels. This meant that, for the first time, steam could be used for propulsion. His engine *Puffing Devil*—the world's first steam locomotive—made its first journey on Christmas Eve 1801.

How steam is created

One of the key innovations pioneered by George Stephenson in his 1829 *Rocket* was the fire-tube boiler, which became a fundamental feature of all steam locomotives. Earlier engines had used a single fire tube to heat the water in the boiler, but Stephenson used 25 copper fire tubes to greatly increase the heat transfer between the firebox and the boiler, meaning that steam could be created much more efficiently. Later engines used superheater elements in place of the fire tubes.

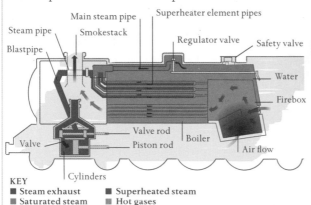

Main steam pipe — Superheater element pipes
Steam pipe — Smokestack
Blastpipe — Regulator valve — Safety valve
Water
Firebox
Valve rod — Boiler
Valve — Piston rod — Air flow
Cylinders

KEY
■ Steam exhaust   ■ Superheated steam
■ Saturated steam   ■ Hot gases

FROM WATER TO MOTION
In a typical steam engine, forward motion is achieved by burning coal to heat water, creating steam that is passed at high pressure to the cylinders in order to turn the locomotive's running gear.

FUELING THE ENGINE
A fireman shovels coal into the firebox of a steam engine, ensuring that the fire tubes are hot enough to superheat the steam in the element pipes.

Steam propulsion

Steam from the boiler is superheated to over 212°F (100°C) and transferred to the cylinders at high pressure, pushing the pistons which turn the driving wheels via a series of pivots and rods, converting linear motion to rotation.

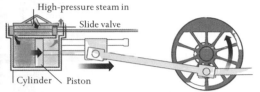

High-pressure steam in
Slide valve
Cylinder — Piston

PHASE 1: OUTWARD STROKE
High-pressure steam is fed via a sliding valve into the front of the cylinder where it expands and pushes the piston, which rotates the wheels by a half-turn.

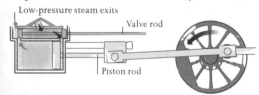

Low-pressure steam exits
Valve rod
Piston rod

PHASE 2: EXHAUST
The wheel is connected to the sliding valve via a series of rods. These open the valve to allow the steam, which has now lost pressure, to escape.

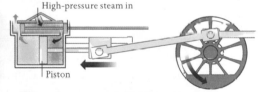

High-pressure steam in
Piston

PHASE 3: RETURN STROKE
The movement of the valve also allows high-pressure steam to enter the back of the cylinder, allowing the return phase of the stroke to begin.

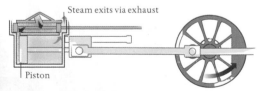

Steam exits via exhaust
Piston

PHASE 4: EXHAUST
Once the wheels have made another half turn, the sliding valve allows spent steam to escape and fresh steam to enter, and the cycle begins again.

The First American Railroads

IN 1828, THE 90-YEAR-OLD CHARLES CARROLL STEPPED UP to make the inaugural speech at the breaking ground ceremony for the new Baltimore and Ohio railroad. Carroll had witnessed the birth of the United States firsthand—he was the only surviving signatory of the American Declaration of Independence. Half a century later, as he commemorated the launch of this ambitious project that aimed to reach into the heartland of the continent, his words proved to be remarkably prescient: "I consider this among the most important acts of my life, second only to my signing the Declaration of Independence, if even it be second to that."

The United States, which had only recently freed itself from the shackles of colonialism, was far behind Great Britain, its former colonial master, in terms of technological development. Its earliest railways—it soon adopted the name "railroads"—were dependent on British imports, as were its river-boats, factories, and mining operations, all of which ran on British steam engines. To catch up, American industrialists kept a close eye on British railway developments, and often traveled across the Atlantic to pick up the latest information. The size of the US, and the ambition of its people, made it fertile ground for the iron road, and it was perhaps inevitable that the new nation would soon boast more miles of track than the rest of the world put together. In fact, the US would end up, at the peak of the railroad boom in 1916, with more than 250,000 miles (400,000km) of line, by far the biggest rail network the world has ever seen.

COLONEL JOHN STEVENS
Inventor and lawyer John Stevens saw the early commercial potential of the railroads over the waterways of the US.

NORTH AMERICAN RAILROADS TO 1860

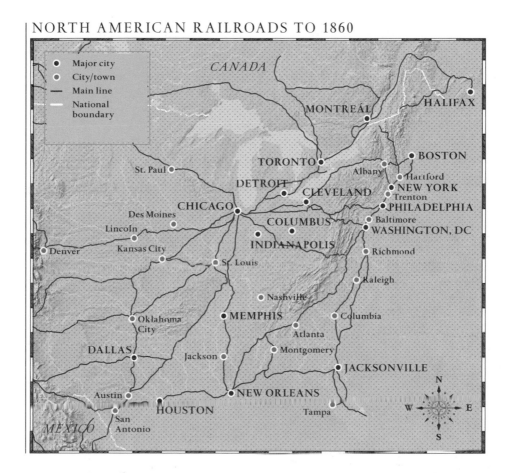

Until the advent of the railroads, transportation in the US had been difficult and slow. There were a few canals, but these iced up in winter. The roads were very poor, owned by turnpike trusts that were unable to maintain them properly as the tolls that were collected were insufficient. Steamboats were the best form of transportation, but they only gave access to certain parts of the country. The first American railroad pioneer was Colonel John Stevens, a successful steamboat designer and operator who was obviously rather taken with the railroads since he wrote a pamphlet entitled *Documents Tending to prove the Superior Advantages of Rail-ways and Steam-Carriages over Canal Navigation*. As early as 1815, he obtained the first railroad charter for permission to construct a line linking the Delaware river near Trenton with the Raritan river in New Jersey, though in the event it was never built—no investors came forward to finance the plan, which was far ahead of its time. Undeterred, in 1825 Stevens designed and built a steam locomotive, which ran on a circular track on a narrow-gauge line at his estate.

Stevens and his two sons were involved in a number of other early projects, notably the Philadelphia and Columbia line, which was built to link the port of Philadelphia with Columbia on the Susquehanna river to give merchants in Philadelphia access to Harrisburg and Western Pennsylvania. They also founded the Camden and Amboy Railroad, which ran from Camden, across the Delaware river from Philadelphia, to Amboy, on the New Jersey shore opposite New York. Initially, all these lines were horse-drawn, but inevitably, given the distances involved, locomotive traction was considered—and for that, British technology was needed. John Stevens' son Robert traveled to Britain and brought back a locomotive, *John Bull*, which was built in the Stephenson Works. It arrived in parts and was assembled by Isaac Dripps, an engineer who fitted pilot wheels at the front to help guide the locomotive around the sharper bends on the American railroad, and who was also credited with inventing the "cowcatcher"—in reality a cow killer that pushed away cattle or deer that had roamed onto the line, invariably fatally.

Building lines in the US in the early 19th century was by no means easy. First, the promoters had to obtain a charter from the state government, then they had to persuade investors, who were often local people, to support the plan, and finally they had to find sufficient workers to build the line as there was often a shortage of labor. There was one key advantage compared with other countries. Once a charter was obtained, the railroad company had "eminent domain"—the right to take over any land required for the line's construction. Sometimes, though, the law was difficult to apply in

"The introduction of so powerful an agent as steam to a carriage on wheels will make a great change in the situation of man"

THOMAS JEFFERSON, 1802

AN EARLY SUCCESS
The pioneering *John Bull* locomotive served for
over 30 years. It is the oldest working self-
propelled vehicle in the world.

practice. When the Erie Railroad was being constructed across
upstate New York, it was planned to cross Native American land.
The local tribe demanded $10,000 (in modern terms, around
$300,000, or £185,000) for the right of way. Appalled, the railroad
works manager blustered that the land was no good for anything
else apart from growing corn or potatoes. The local chief responded:
"It pretty good for railroad," and got the money.

Most of the early railroad development was stimulated by
competition between the great cities of the eastern US, such as
Baltimore, Philadelphia, New York, and Boston. Each wanted to obtain
cheap access to the Midwest, where towns were growing rapidly,
creating an important market for produce. Baltimore proved to be the
most adventurous in promoting a railroad stretching deep into
the hinterland. The Baltimore and Ohio was the most significant of
these early schemes, being the first attempt to build a rail link between
an Atlantic port and the Ohio River, and so reaching the Midwest.

As with so many of these early lines, the promoters of the Baltimore
and Ohio were unsure about whether to use horses or steam
locomotives to haul the trains. Given that they wanted the line to
reach the town of Wheeling, nearly 400 miles (650km) from Baltimore,
it is extraordinary that they even considered using equine power, but
they arranged a competition between the hay eater and the coal
burner. A locomotive builder, Peter Cooper, had built a little engine
nicknamed *Tom Thumb* for the line, and it proceeded to impress the
promoters with a test run on the initial 13 miles (21km) of track that

had been completed, reaching an exhilarating 18mph (29kph). On the run back toward Baltimore, Cooper foolishly agreed to race his locomotive against a powerful grey horse. The animal soon took the lead, thanks to its faster acceleration, but was then overtaken by the steady little engine when Cooper opened the safety valve to provide extra power. However, he overreached himself: after the locomotive had gained a significant lead, the belt that drove its pulley snapped, and the engine eased to a halt. The equine victory proved Pyrrhic, however, as Cooper had done enough to persuade the promoters that steam haulage, rather than horsepower, was the only way to make the line viable. Although work started in 1828, and trains started operating on part of the line two years later, it was not until 1853 that the tracks reached Wheeling on the Ohio River owing to legal, financial, and technical difficulties.

THE RACE BETWEEN OLD AND NEW
The *Tom Thumb*'s race against a horse became the stuff of railroad legend. The locomotive was the first American steam engine to run on a commercial railroad.

Further south, there was a far longer pioneering line, which was completed much more quickly and used American technology. The Charleston and Hamburg was an attempt to revive Charleston's foreign exports, which had gone into decline, and its local merchants hoped to secure the trade of the rich cotton-growing area in the region. They chose steam power from the outset, and the first engine, the *Best Friend of Charleston*, built at the West Point Foundry in New York, pulled its first train in December 1830. Unfortunately, a couple of months later, the pioneering locomotive suffered an untimely demise when an inexperienced fireman, annoyed at the sound made by the escape of steam from its safety valve, sat on the offending piece of machinery— which caused the boiler to explode, killing the fireman and scalding the driver. Despite this mishap, the line was complete by 1833, and at 136 miles (219km), it was, for a time, the longest in the world.

American railroads differed from their European counterparts in several respects. The key difference was one of scale, not just in the extent of their reach, as they gradually extended further and further west, but also in the size of the trains and locomotives themselves.

RAILROAD COACH
This replica of one of the earliest Baltimore and Ohio Railroad coaches shows how they were little more than stage-coach carriages placed on railroad trucks.

FRONT   BACK

B&O CENTENARY MEDAL, 1827-1927
This medal was issued to commemorate 100 years
of the history of one of America's oldest railroads,
the Baltimore and Ohio. It depicts the famous early
locomotive *Tom Thumb*.

Their characteristically huge, bulbous chimneys—needed to contain
the sparks that might otherwise set fire to the countryside—were far
taller than their European equivalents, for the American lines had
few bridges or tunnels. Consequently, even today American trains
are almost 3ft (1m) taller than those in Europe, enabling them to
carry much greater loads. Overall, US railroads were bigger in every
sense than those across the Atlantic. They covered greater distances,
and were longer and heavier because they used stronger and larger
locomotives, all of which gave them a distinctive style (see pp.40–41).

The efforts to cut the costs of the new lines were successful, and
US railroads were far cheaper to build than their European equivalents,
but as a result, they were also less reliable and slower. Some aspects of
the railroads were, however, better from the start. Locomotives, for
example, were fitted with cabs for the crew, a "luxury" that was
necessitated by the rigors of the US climate, but which did not become
universal elsewhere until much later. Right from the start, too,
passengers traveled in carriages that were open plan, rather than in
individual compartments like those in Europe. These were necessary
because traveling longer distances meant that travelers required ready
access to conveniences, a facility that was not available on most
European trains until well into the second half of the 19th century.

These early lines were successful and mostly profitable, which attracted a wave of investment into the new industry. By 1837, at least 200 railroads were being promoted. Many of these were unrealistic or promoted by crooks intent on cheating potential investors, but many plans were still completed, and by the end of the decade 2,750 miles (4,425km) of railroad were in operation—a remarkable rate of progress. The spur for most of these lines was freight, particularly coal and minerals, but increasing numbers of passengers also flocked to the trains. Soon, the short lines were followed by long trunk railroads such as the Erie and the Pennsylvania lines, linking the Eastern Seaboard with the Midwest. Later, in the second half of the 19th century, the transcontinentals brought the railroads to the West. Before long, every town wanted to be connected to the railroad network as it was considered vital for their prosperity. Prominent local citizens would band together and form a company to obtain a charter, often investing their own money. A mere 20 years later, at the outbreak of the Civil War (see pp.60–65), there were nearly 30,000 miles (48,280km) of railroad in the US.

The railroads, in fact, grew symbiotically with the US economy, transforming the nation from a predominantly agricultural country into the industrial powerhouse of the world, all within a few decades. It is impossible to know whether the tracks spread so quickly because of the rapid growth in wealth, or whether it was the other way around, but there is no doubt that the US thrived because of the growing railroad system and that the railroads welded this vast nation together (see also pp.120–27).

The Early Years of American Steam

Starting with the Baltimore and Ohio Railroad in 1828, the growth of the US railroad network enabled the Industrial Revolution to take hold, and opened up the west coast to exploration and colonization. American locomotive design engineers produced some innovative designs, including geared locomotives and the first rack railroad (see pp.108–109).

BALTIMORE AND OHIO GRASSHOPPER NO.2: *ATLANTIC* (1832)
Named because its long connecting rods and vertical cylinders gave it an insect-like appearance when moving, the Grasshopper *Atlantic* was the first US-designed locomotive to see commercial service on American soil.

BALTIMORE AND OHIO GRASSHOPPER NO.8: *JOHN HANCOCK* (1836)
The *John Hancock* was an improved model of the Grasshopper class, featuring a covered cab and dual-powered axles. It was used continuously on the Baltimore and Ohio Railroad from 1836 until 1892.

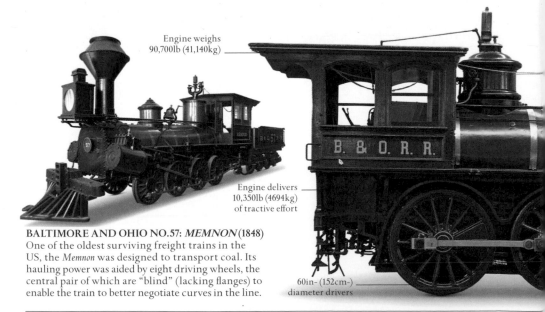

Engine weighs 90,700lb (41,140kg)

Engine delivers 10,350lb (4694kg) of tractive effort

60in- (152cm-) diameter drivers

BALTIMORE AND OHIO NO.57: *MEMNON* (1848)
One of the oldest surviving freight trains in the US, the *Memnon* was designed to transport coal. Its hauling power was aided by eight driving wheels, the central pair of which are "blind" (lacking flanges) to enable the train to better negotiate curves in the line.

CUMBERLAND VALLEY RAILROAD NO.13: *PIONEER* (1851)

The *Pioneer* was a lightweight locomotive with only one pair of driving wheels. It was used to haul up to three passenger cars, and is pictured here in the livery of its retirement in 1901.

SHAY NO.1: *LEETONIA* (1906)

The *Leetonia*, one of engineer Ephraim Shay's geared locomotives, used toothed driveshafts driven by three vertical cylinders to power all 12 of its wheels, making it ideal for hauling lumber at low speeds on steep grades and tight curves.

MOORE-KEPPEL AND COMPANY CLIMAX NO.4 (1913)

Similar to the Shays, the Climax-class geared engines were used by lumber companies. A cylinder on either side of the boiler powered eight wheels via a centrally mounted transmission.

Conical smokestack

READING COMPANY NO.1251 (1918)

Built using parts from older locomotives, No.1251 was a switching engine with six driving wheels. Its "saddletank" design meant that no tender was required—water was carried in a tank on top of the boiler.

Boiler runs at a pressure of 75psi

Front driver was originally flangeless

BALTIMORE AND OHIO NO.147: *THATCHER PERKINS* (1863)

Named after its designer, the company's Master of Machinery, *Thatcher Perkins* was rushed into service to meet the demands of the Civil War. Its six driving wheels and four leading wheels were designed to cope with the steep tracks in the Appalachian Mountains.

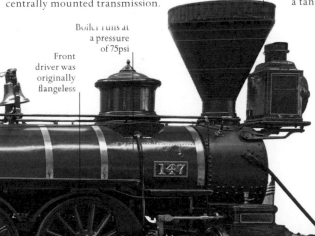

19 × 26in (48 × 66cm) cylinders

Europe Takes to the Rails

A S BRITAIN WAS PIONEERING THE RAILROADS, the cafés of Paris were abuzz with news of the recent developments – but not everyone was enthusiastic. Just as railroad opponents in Britain had warned that it might be impossible to breathe when traveling at more than around 30mph (50kph), or that cows would stop producing milk because of the noise of the railroad, in France and the rest of Europe, too, there were eminent doubters of this new technology. Indeed, the French habit of philosophizing over major issues meant that Parisian intellectuals discussed the pros and cons of the new railroads in great detail—the writer Edmond de Goncourt warned that when traveling on the railroad, "one was so jolted about that it was quite impossible to collect one's thoughts."

However, despite the naysayers, the major European countries—particularly France, Belgium, and the German Confederation—were now scrambling to construct their own lines. The economic advantages of doing so were obvious, but there were political reasons too—railroads were not only vital assets in war, they could help bind nations together. At first, Europe was dependent on Britain both for technology and drivers, but many countries soon became independent, notably France, whose early efforts were close behind Britain's. In 1823, during a brief revival of the French monarchy, Louis XVIII signed an Act that permitted the construction of France's first railroad. The 14-mile (23-km) track ran between St.-Étienne and Andrézieux in the Massif Central and was built to carry coal from the

MARC SEGUIN
French inventor Marc Seguin (1786–1875) pioneered the multitube boiler for steam engines. He also developed the first suspension bridge in Europe.

mines to the Loire for shipment to the rest of the country. The line opened in 1827 and, although horse-drawn, it was an instant success. As a result, extensions were added, and by 1832 the line, which now used locomotives and carried passengers, stretched to the major city of Lyon. The elaborate French cars were an improvement on the rather more austere British trains and were divided into compartments, an arrangement that soon became standard across Europe.

France's equivalent of George Stephenson (see pp.22–29) was Marc Seguin, a scientist and inventor who had in fact advised Stephenson on how to improve the boilers of his locomotives. He produced two locomotives for the extended St.-Étienne–Lyon railroad, each featuring a multitube boiler (see pp.30–31) and a mechanical fan to deliver oxygen to the fire. Later, in another case of Anglo-French cooperation, Robert Stephenson built locomotives designed by Seguin. The lengthy debate over the advantages and disadvantages of railroad construction slowed the pace of development in France, so there were only 350 miles (560km) of track by 1840, compared with 2,000 miles (3,200km) in Britain.

The cause of the railroad promoters was not helped when the world's first major rail disaster occurred between Versailles and Paris, in May 1842. The train, returning from Versailles, was so heavily laden with vacationers who had been watching the celebrations for the king's fête that it required two locomotives to haul it. The leading engine suffered a broken axle—a relatively commonplace event in the early days of railroads—and then derailed, along with three passenger cars, which quickly burst into flames. The death toll was at least 50, and may have been as many as 200; people couldn't escape because they were locked in and many bodies were consumed by the inferno. As a result, the French authorities stopped locking passengers into their cars, although the practice continued elsewhere, contributing to the high death toll of the 1889 Armagh disaster, the worst in Irish history (see pp.139).

Many other countries around Europe joined the Railway Age in the 1830s, and the state was usually much more directly involved than in Britain, where the government had remained aloof. Nowhere was this more true than in Belgium, a new country carved out of the Netherlands in 1830 and anxious to demonstrate its independence. The railroads were an ideal way of doing so, since building a railroad system was thought to stimulate a sense of national identity. Consequently, the country's first king, Leopold I, approved the design of a whole network, and in 1834 building started on the first line. This crossed the entire country from Antwerp in the north to Mons in the west, and into Prussia via Aachen in the east—a total of 154 miles (248km), a very ambitious plan at that time. Together with an Ostend–Liege line forming an east–west axis, Belgium quickly created a fully planned national railroad system—the first of its kind in the world. Inevitably, George Stephenson was involved, his company providing the first three locomotives and he himself traveling incognito in 1835 on the first train carrying the royal party. When the train broke down, Stephenson went to the engine to help fix the problem, and was knighted by the king for his pains. The support of the government gave Belgium a lead in railroad development, and by 1843 most of the heart of Belgium's railroad network—which forms a "cross" shape centered on Brussels—had been built, giving the heavily industrialized country the densest network in the world.

Germany's first railroad also opened in 1835, doing so in Bavaria, where King Ludwig approved a steam-hauled line that ran the 4 miles (6.5km) between Nuremberg and Fürth. Unlike most of the inaugural lines, this was built mainly for passengers, as it relieved traffic on the busiest highway in Bavaria between the two towns. The congestion was the result of peculiar local circumstances. For centuries, the Nuremberg authorities had forbidden laborers and foreigners to live in the town, so they had to commute from Fürth, which had become a dormitory town, presaging a use for the railroad that remains commonplace today. Saxony, another important German state, followed Bavaria's lead, building Germany's first major railroad, linking Leipzig with Dresden. Saxony was the industrial heartland of

NUMBER OF LOCOMOTIVES IN GERMANY BY 1880

9,400

Germany, similar in character to the northwest of England, where the pioneering Liverpool and Manchester line had been built (see pp.25–29). More than 200 factories had sprung up in the region, and local industrialists, realizing that a railroad was essential to carry the minerals

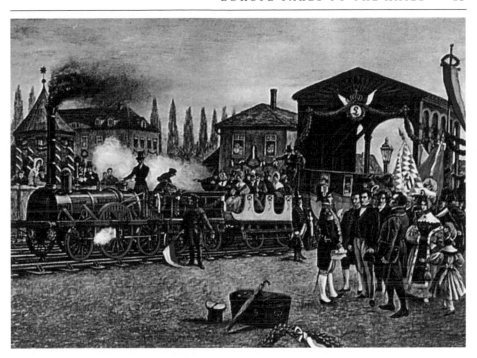

THE FIRST GERMAN RAILROAD
This contemporary painting shows the departure
of the locomotive *Adler* from Nuremberg in 1835.
The *Adler* was the first commercial locomotive in
use in Germany and was in service until the 1850s.

and ore needed by the factories, quickly raised the money to build the
65-mile (105-km) line, which was built remarkably swiftly thanks to
the use of British technology and personnel.

The railroads were particularly important for Germany as a
means of uniting the country. As early as 1817, economist and
visionary thinker Frederick List had understood the importance of
the railroads for Germany. He argued that a nation could prosper
only through trade and industry, and that a fast, efficient rail network
could carry food and industrial products throughout the country.
His theory was borne out. Customs duties between states were
soon scrapped as impractical, and within a generation of the first rail
line being completed, Germany's railroad system had helped it
become a powerful unified state.

Like the Nuremberg–Fürth line, the first line in Holland, another
relatively early starter, was also built for passengers rather than
freight. The Holland Iron Railway between Amsterdam and
Haarlem opened in 1839 and reached Rotterdam eight years later. It

was extremely successful, killing off the competing coach and barge traffic that had previously dominated the route between the country's two main cities. Holland, though, did not have the centrally planned scheme of its Belgian neighbor and so never developed such an intensive network.

The Belgian, German, and Dutch railroads were all steam-hauled, but the idea that railroads could be operated successfully by horses still lingered. In Austria, remarkably, the early network was entirely horse-drawn, and grew to an extensive size. It included the world's longest horse-drawn railroad—a 90-mile (145-km) line linking Linz in Upper Austria with Budweis in Bohemia (home of the world-renowned beer, now in the Czech Republic). The horse-drawn system was later extended further to the salt works at the Upper Austrian health resort of Gmunden, and by 1836 it had reached an impressive 170 miles (274km). Only then did the Austrians replace horses with steam engines.

The railroads of both Italy and Russia began as royal initiatives, albeit for the private purposes of connecting royal palaces. In Italy, the King of the Two Sicilies, the portly Ferdinand II, fancied connecting his main palace in Naples to his other residence on the bay of Naples at Portici with a 5-mile (8-km) railroad line, which was completed in October 1839. The idea had been recommended to him by a Frenchman, Armand Bayard de la Vingtrie, who was keen to make money from railroad plans. The first section of the line was built quickly and proved to be immensely popular, carrying up to 1,000 people per day. Oddly, the king himself eschewed traveling on the inaugural journey, perhaps because he was aware of the risks of train travel.

In Russia, Tsar Nicholas I built the first railroad line between his main residence in St. Petersburg and Tsarskoe Selo, the enormous royal residence favored by Catherine the Great. He was advised by an Austrian engineer, Franz Anton von Gerstner, who

THE RAILROADS COME TO ITALY
This 1839 painting by Salvatore Fergola depicts the inaugural train leaving Portici station in Naples, Italy. The line was Italy's first railroad, and today forms part of the Naples–Salerno line.

was keen to bring Russia into the railroad fold, and had sought to build a far more ambitious line between St. Petersburg and Moscow. Initially, however, only the Tsarskoe Selo line was commissioned, and opened in 1837 with the first train, carrying eight full passenger cars, taking a mere 28 minutes to reach Tsarskoe Selo—an impressive average speed of almost 30mph (50kph). The following year, the line was extended by 16 miles (26km) to Pavlosk, a small holiday resort complete with buffets, concerts, and a ballroom to entertain the St. Petersburg crowds on their day trips. To attract the crowds, the railroad subsidised the public entertainment at Pavlovsk, which is described in Dostoevsky's *The Idiot* as "one of the fashionable summer resorts near St. Petersburg." At first, the line was hauled by horses and various locomotives imported from Britain and Belgium, but before long the animals, exhausted from pulling the heavy trains, were put out to grass and steam locomotives were introduced. The line was a great success, with people flocking to the railroad both out of curiosity and the desire to sample the attractions. More than 725,000 people traveled on the line in its first year, an average of

THE LEGACY OF TSAR NICHOLAS I
The St. Petersburg–Moscow line was Russia's first major railroad. Wary officials warned the tsar—wrongly, as it turned out—that allowing people to travel on the line would result in an uprising.

"Railway! A magical aura already surrounds the word; it is a synonym for civilization, progress, and fraternity"

PIERRE LAROUSSE, 1867

2,000 per day. The 400-mile (640-km) long St. Petersburg–Moscow line was completed in 1851, becoming one of the longest and most impressive trunk routes in the world at the time.

Most countries in Europe adopted the 4ft 8½ in (1,435 mm) gauge devised by Stephenson (see p.25), and this proved vitally important in ensuring that a continental network was created, with trains passing through borders relatively easily—although customs and technical factors relating to signaling and driver training usually meant delays at the frontier. Russia and Spain were exceptions— both used a 5-ft (1½-m) gauge for fear of invasion from hostile neighbors, reckoning that making trains change gauge would prove a useful defensive barrier.

Traveling on these early lines was not always made easy, either by the regimes of the day or the railroad companies. Governments were suspicious of their citizens' desire to travel, and the companies were diligent in collecting fares. On the Leipzig–Dresden line, for example, there were no advance ticket sales and access to the station was allowed only a quarter of an hour before the train left—a practice that persists in remote parts of Europe even today. Passengers were required to purchase a return ticket and children under 12 were banned. In Russia, there was also an extraordinary amount of bureaucracy involved in taking a train. Passengers had to obtain an internal passport before traveling, and then go to their local police station to obtain permission for the journey. Even today, Russians have to show their internal passport before buying a ticket for a long-distance journey. Nevertheless, despite the bureaucracy and the technical problems incurred on many pioneering lines, all these early railroads proved popular and successful, spurring the rapid spread of rail travel throughout Europe and to many other parts of the world.

Western European Railroads

In the mid-1830s, railroad technology spread from its birthplace in Britain to continental Europe. Belgium enthusiastically adopted steam locomotion as a means to cement its fledgling nationhood. France stifled the iron road with bureaucracy (see box, opposite), while the German states used the railroads to further the case for national unification. The lines spread in the ensuing decades until Europe was crisscrossed with a vast web of railroads. Spain and Portugal, however, chose to use different gauges from the rest of the continent. This map shows modern Europe's main lines.

SWEDEN

NORWAY

OSLO

STOCKHOLM

GLASGOW   EDINBURGH

EIRE

DUBLIN

UNITED KINGDOM

DENMARK

COPENHAGEN

LONDON   AMSTERDAM   HAMBURG

BRUSSELS   GERMANY   BERLIN

BELGIUM   FRANKFURT   POLAND

PARIS

LUXEMBURG

PRAGUE

ATLANTIC OCEAN

FRANCE

VIENNA

AUSTRIA

BUDAPEST

BAY OF BISCAY

LYON   SWITZERLAND   MILAN   HUNGARY

SLOVENIA

TURIN   ZAGREB

BOSNIA

SARAJEVO

PORTUGAL

MARSEILLE

CROATIA

LISBON   MADRID   SPAIN   BARCELONA

ROME   ITALY

MEDITERRANEAN SEA

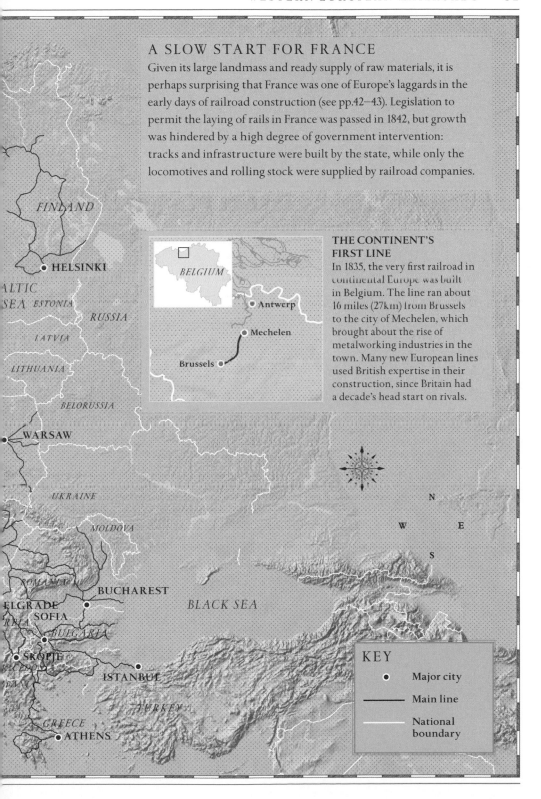

A SLOW START FOR FRANCE

Given its large landmass and ready supply of raw materials, it is perhaps surprising that France was one of Europe's laggards in the early days of railroad construction (see pp.42–43). Legislation to permit the laying of rails in France was passed in 1842, but growth was hindered by a high degree of government intervention: tracks and infrastructure were built by the state, while only the locomotives and rolling stock were supplied by railroad companies.

FINLAND

HELSINKI

BALTIC SEA ESTONIA

BELGIUM

Antwerp

RUSSIA

LATVIA

Mechelen

LITHUANIA

Brussels

THE CONTINENT'S FIRST LINE

In 1835, the very first railroad in continental Europe was built in Belgium. The line ran about 16 miles (27km) from Brussels to the city of Mechelen, which brought about the rise of metalworking industries in the town. Many new European lines used British expertise in their construction, since Britain had a decade's head start on rivals.

BELORUSSIA

WARSAW

UKRAINE

MOLDOVA

N

W   E

S

ROMANIA

BUCHAREST

BLACK SEA

BELGRADE

SOFIA

RBIA

BULGARIA

SKOPJE

ISTANBUL

BANI

TURKEY

GREECE

ATHENS

KEY

- ● Major city
- ── Main line
- ── National boundary

Railroad Mania

THE SUCCESS OF THE FIRST RAILROADS stimulated enormous interest from potential investors and promoters of new lines across the world. There was a tendency for people to try to jump on the bandwagon, creating periods of railroad mania, which occurred in nearly every country that developed a network. Unfortunately, not all of these promoters were honest, and some had the sole intention of trying to make people part with their money.

Oddly, the first mania happened before a single line was completed. In 1824–25, during the run-up to the opening of the Stockton and Darlington Railway (see pp.24–25), entrepreneurs around Britain put forward ideas for other lines, and prospectuses for around 70 were published—a remarkable number, given that they were still an unproven technology. However, a downturn in the economy following a poor harvest in 1825 and a spate of banking failures soon soured the optimism, and almost all of these schemes were quietly forgotten. The success of the Liverpool and Manchester line (see pp.25–29) prompted another spate of railroad promotion in the mid-1830s, but it was the huge surge in applications for new lines from 1844 onward that led to the use of the term "railroad mania." The phenomenon recalled previous promotional fevers such as the South Sea Bubble of 1719–20 and the canal mania of 1791–94, and foreshadowed more recent frenzies such as the dot-com boom of the 1990s.

In Britain, by the mid-1840s, railroad construction was established as a legitimate and profitable business, and the healthy economic conditions were ripe for a major boom. The usual method for raising capital was to organize local public meetings at which investors would buy "scrip"—vouchers that could later be exchanged for future shares—for a small deposit in the proposed company. Having raised part of the capital, the promoters would then go to Parliament for approval of their proposal. There had been something of an economic slump in the early 1840s, but as the economy recovered there was an upsurge in railroad promotion. Railroads were seen as a method of getting rich quickly, and from 1844 to 1847, British parliamentary approval was obtained for more than 8,000 miles (12,800km) of line—nearly five times the amount that had already been laid, and a large proportion of Britain's current rail network of 9,790 miles

(15,750km). At the time, building a railroad was a relatively simple task, provided the terrain was favorable—no more difficult than opening a local store or erecting a row of houses. Once permission was obtained, it was a matter of creating a narrow permanent right of way, laying the track, and putting in the odd raised wooden platform at stations. There was none of the complexity of today's technically sophisticated railroads.

The rapidity with which the mania took hold was astonishing. In 1843, only 100 miles (160km) of new line were approved by the British Parliament, but over the following three years the figures were respectively around 800 miles (1,200km), nearly 3,000 miles (4,800km) and an astonishing 4,500 miles (7,200km), as 272 acts were granted. The scale of this boom can be judged by the fact that it represented a theoretical amount of £700m worth of capital investment—ten times

"THE RAILROAD JUGGERNAUT OF 1845"
This cartoon depicts the ruthless business of railroad mania with vultures circling overhead, crocodiles wearing lawyers' wigs, and the devil perched on a locomotive called *Speculation*.

Britain's annual exports at the time. However, the boom collapsed even faster than it grew. The economy slumped again in 1847, and as a result, only 17 miles (27km) of new line were approved by 1849, but an astonishing two thirds of the mileage approved thus far was eventually completed within a few years, and even some of the failed projects were revived in the smaller booms of 1852–53 and the 1860s.

Although most of the projects, even the failed ones, were genuine attempts at building a railroad, some of them were blatantly fraudulent, while others were based on the belief that the expansion would last forever. This was the how the biggest fraudster of the period, George Hudson, created a massive empire, which inevitably collapsed when the bubble burst. A strange-looking fellow, who one biographer describes as having "a cannon-ball of a head set upon his bulky shoulders, the formality of a neck having been disposed of," Hudson was full of energy and had a fondness for the good life— perhaps the typical profile of a fraudster. He did, in fact, succeed in building several railroads, including the core of the Midland Railway, which was one of the most extensive lines in Britain. He had some other good ideas, too, such as establishing a ticket clearing house through which railroad companies reimbursed each other for running trains on each other's tracks. As he grew in confidence, however, he used the money he raised for new schemes to pay dividends on previous projects—what is now called a Ponzi scheme. His accounting practices had, in fact, been so dubious that on his demise it was impossible to find out where all the money had gone. He had become rich, as one wag put it, "by keeping everything but his accounts." His career, so glittering that he was elected Lord Mayor of York and later an MP, came to an abrupt end in 1849 when his dishonesty was exposed, after which he vanished into obscurity.

THE INFAMOUS GEORGE HUDSON
Railroad promoter, entrepreneur, and fraudster George Hudson created a business empire that forever changed the face of railroad development.

"This big swollen gambler... deserved a coalshaft from his brother mortals"

THOMAS CARLYLE, PHILOSOPHER AND WRITER, ON GEORGE HUDSON

Switzerland underwent a period of railroad mania in the 1850s, helped by Alfred Escher, a powerful Swiss businessman and politician, and a keen supporter of the railroads. He had observed the growth of railroads in other countries, and feared that Switzerland was missing out on the economic prosperity offered by a rail network and would become "the sad face of Europe's backwater." In 1852, he helped push through a law allowing private companies to build and run railroads. This launched a frenzy of construction by competing companies—joined by Escher himself, who founded Swiss Northeastern Railway and enjoyed great success. Over the following decades, the Swiss rail network extended across the country, including the ambitious Gotthard Railway in the 1870s (see pp.105–106).

In France, a law passed in 1865 to encourage railroad construction in remote areas led to a period of rapid railroad expansion. At the time, France's railroads were controlled by six large companies, and the idea was to encourage new entrants on to the network to ensure better coverage of rural areas where the six were reluctant to go. The law allowed local authorities to sponsor these lines, and speculators piled in, seeing the possibility of making money out of the grants being offered, despite the fact that most of these lines would never be viable. Within ten years, nearly 3,000 miles (4,800km) of remote lines had been built, many with a narrow gauge to reduce costs. However, few of them made money and they soon had to be rescued by the state, forming the core of what later became the French nationalized railroad system, SNCF.

In Italy, too, it was the government that stimulated a period of rapid expansion. Creating a national rail network was seen as essential after the unification of Italy in 1861, and the state sponsored a rapid expansion through a concession system. However, most of the companies that built the lines had insufficient capital and got into financial difficulties when the lines, which were poorly constructed, failed to attract enough passengers. The state inevitably took over and nationalized the railroads, which resulted in many investors losing money.

Perhaps unsurprisingly, it was the US that suffered the biggest bout of "railroad fever," as they called it. There were, in fact, sporadic boom periods of railroad construction throughout the latter part of the 19th century, as humorously described by the historian Stewart H. Holbrook, who related the story of a fictional town called Brownsville:

> First some up and coming individual, or simply a fanatical dreamer, said forcibly that his home town Brownsville needed a steam railroad… the idea grew and blossomed and burgeoned and even soared meanwhile taking on all of the beautiful hues of the sky in the Land of Opportunity. It also dripped with gold, gold for all of Brownsville, soon to be a mighty metropolis, teeming with commerce.

… and so on, until an application was made to the state for permission to build the Brownsville Railroad. This type of scene was enacted many times over across the US during every bout of railroad fever.

The US has more than its share of crooks in its railroad history. A group of "robber barons" emerged from the early railroad companies, often using dishonest methods of speculation to gain control of profitable lines. It was a time of wild risks and gambles. In one famous incident—a takeover battle for the Erie Railroad—three railroad barons, led by Jay Gould, holed up in a hotel in New Jersey with several million dollars in cash, protected by armed guards to evade the jurisdiction of the New York courts, which had found in favor of their rival, Cornelius Vanderbilt. (Gould and his associates did eventually win the battle for the Erie Railroad – see pp.241–42). Such events, while not commonplace, were part and parcel of the colorful period of railroad speculation in the US.

The final bout of speculative building in the US was in the early 20th century, and involved a series of tramways called "electric interurbans." These were a hybrid of trains and trams that connected towns up to 50 miles (80km) apart with cheaply built single lines sited next to existing highways. There was a remarkable period of expansion of these lines from just 2,000 miles (3,200km) at the turn of the century—by 1906 there were 9,000 miles (14,400km), and by the outbreak of World War I, 15,000 miles (24,000km). By then, it was possible to travel all the way from Wisconsin to New York using a series of interurbans. It was a cheap ride, costing 10 cents for the journey, but it was slow, for the interurbans averaged around 20mph

"NORD 2-3-0"
A French locomotive leaves a depot in Boulogne.
France's national rail system, the SNCF, was founded to
rescue the struggling private railroad lines following a
surge in development. It still operates today.

(32kph) at best. Sadly, according to one historian, "the interurbans
were a rare example of an industry that never enjoyed a prolonged
period of prosperity," and most investors lost all their money. The
demise of the interurbans was swift because they were inherently
unprofitable, serving sparsely populated areas and facing competition
from the growing use of motor vehicles. Already at the start of World
War I, systems were closing and they were all but wiped out by the
1930s, when their vehicles and tracks needed renewing and there was
no money for such investment.

All the manias across the world left their mark. Many investors
lost their shirts, but many of the lines were built and a good proportion
of those survive today. As with other industries, the manias had
their roots in genuine need, and many of today's lines worldwide
owe their existence to these excesses.

Wheels and Trucks

The wheels of a locomotive are mounted on a chassis known as a truck (see box, below) and are designed to keep the engine aligned with the tracks. Each wheel tapers slightly from the inside outward, which helps to steer the train around curves (see panel, opposite), and is fitted with a projecting rim or "flange" on the inside edge. Ordinarily this should not touch the track, and is a safety feature to prevent the train from derailing. Wheel sets—two wheels joined by an axle—are variously sized and perform subtly different functions: large driving wheels are powered by the pistons of the locomotive, while smaller, unpowered leading and trailing wheel sets support the weight of the engine and enable the train to pass through junctions and bends in the line.

Trucks

Each leading and trailing wheel set is mounted on a frame beneath the car. The strength and rigidity of the structure, or "truck," enables the wheel sets to resist torsional forces when the train turns. Most trucks fix wheel sets in place in an inflexible frame, but "steerable" trucks allow the axles to rotate laterally around a pivot between the two wheel sets, increasing the stability of the train around bends. Modern trucks also house the train's braking and suspension systems.

Each wheel set consists of two wheels connected by an axle

Axles may rotate laterally for steering

GERMAN RAILROAD TRUCK, c. LATE 19TH CENTURY

THE IRON GIANT
A steam engine is lowered onto its truck, with large driving wheels and smaller leading wheels visible. As well as being strong enough to carry heavy engines, train wheels and trucks were engineered to withstand immense rotational and torsional forces.

How it works

The flanged wheel was invented in 1789 by English engineer William Jessop. The raised rim on the inner wheel edge prevents derailment and does not touch the rail during normal running, unless the track is poorly maintained. The conical edges of the train wheels allow the wheel sets to slide across the heads (tops) of the rails, enabling the train to follow curves. Engineers observed that the characteristic side-to-side swaying action of a train in motion was due to its tapered wheel sets wobbling up and down the railheads in order to "hunt" for equilibrium. They termed this movement "hunting oscillation."

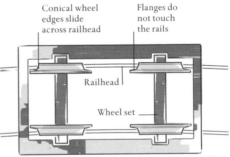

Conical wheel edges slide across railhead

Flanges do not touch the rails

Railhead

Wheel set

TOP VIEW

NEGOTIATING CURVES

On a curved track, the train's outer wheels have to travel slightly further. To compensate, the wheel set slides across the railhead, allowing the outer wheels to use the larger radius of their inner edge. The inside wheels meanwhile slide onto the smaller radius of their outer edge. This action allows the train to lean into the bend, much like a cyclist leaning into a corner.

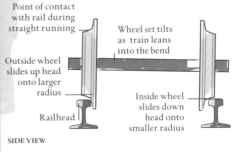

Point of contact with rail during straight running

Wheel set tilts as train leans into the bend

Outside wheel slides up head onto larger radius

Railhead

Inside wheel slides down head onto smaller radius

SIDE VIEW

The American Civil War

BOTH GOVERNMENTS AND MILITARY commanders soon appreciated the potential of the railroads for warfare. Shortly after its opening in 1830, the Liverpool and Manchester Railway (see p.25–29) was used to carry a regiment of troops from Manchester to the docks in Liverpool en route to quell a rebellion in Ireland. The 31-mile (50-km) journey took just over two hours, rather than the two days it would have taken by foot, and the troops arrived in a much fresher state. Around this time, revolutionary fervor was brewing in Europe, and by the late 1840s, rulers across the continent were using the railroads to help crush these revolts. The first major movement of troops by rail took place in 1846, when a contingent of 14,500 Prussian soldiers was sent to put down the Krakow rebellion of Polish nationalists against their Austrian rulers. The Prussians rapidly suppressed the uprising, with considerable bloodshed, after covering the 200-mile (340-km) journey in just two days. Then, two years later, Tsar Nicholas I of Russia, the most reactionary of the mid-19th-century monarchs, sent 30,000 troops on the newly built Warsaw–Vienna railroad to help his ally, Emperor Ferdinand I of Austria, to put down another nationalist rebellion—this time in Hungary. Again, the result was a defeat for the revolutionaries, with much loss of life.

After this, the scale of troop movements around Europe increased. In the winter of 1850, Austria used the railroads to send a 75,000-strong army, along with 8,000 horses and untold equipment, from Vienna to Bohemia. Due to poor weather and the fact that trains traveled along a single line, the move took longer than expected—26 days for a mere 150-mile (240-km) journey – demonstrating that an enormous amount of planning was still needed to use the railroads effectively. Three years later, the French organized a massive movement of troops during the Crimean War, in what is now Ukraine. Most of the army of 400,000 men despatched to fight in the Crimea traveled to the Mediterranean seaports on the railroad being built between Paris and Marseilles. Although the line was not finished, troops were able to use long sections of it, greatly improving their journey. In fact, it was during the Crimean War that the first railroad line intended specifically for military purposes was built. This was the work of the British, who fought alongside the French, and who had struggled to bring men and supplies up to Sevastopol, which they were besieging, from the port of Balaklava 8 miles (13km) away. The hilly road was little more than a dirt track, and a bottleneck soon built up in the port. To relieve it, a group of "navvies" (see pp.84–89) was sent from Britain to build a line between the port and the camp outside Sevastopol. They were a wild bunch, causing mayhem on their journey by getting drunk and trying to catch apes in Gibraltar, but they were very effective builders. The Grand Crimean Central Railway—an overblown name for a short, narrow-gauge railroad powered partly by horses and partly by steam engines—was built remarkably fast (just seven weeks) in 1855. It proved to be of enormous value, enabling troops and provisions, including heavy guns, to be hauled up the hill to support the assault on Sevastopol. The railroad made it possible to bring an unprecedented number of guns to bear on the town, which eventually succumbed under the barrage, effectively ending the war.

It was during the American Civil War, however, that railroads came of age as strategic assets. The conflict had been brewing for a long time, and had its roots in the differences between the northern and southern states. The North was industrializing and developing its economy on the basis of manufacturing, but the South remained primarily agricultural (cotton was its principal export) and depended on slave

WAR ON THE RAILS
This newspaper sketch shows the capture of a train near Gunpowder Ridge, on July 11, 1864, during the invasion of Maryland.

DERAILED CIVIL WAR LOCOMOTIVE
Soldiers survey the aftermath of an attack in 1864.
Both the deployment and destruction of railroads
played a crucial role in the American Civil War.

labor. The trigger for the war was the inauguration of Abraham Lincoln as president in 1861. Lincoln opposed the expansion of slavery to new states; the southern states, fearing the complete abolition of slavery, seceded from the Union in the spring of 1861, thus starting the war.

The Civil War was a bloody affair, claiming the lives of 620,000 soldiers. The battles between the Southern Confederates and the Northern Federal Army were fought on an unprecedented scale. Throughout history, even wars that raged over long periods typically featured only a handful of battles. However, during the four years of the War Between the States, a remarkable 10,000 military encounters took place, of which nearly 400 were significant enough to be considered full-scale battles— which means a battle was fought every four days. Moreover, the war was fought over an area even bigger than Europe, a vast territory that was only made accessible by the railroads.

By the outbreak of the war, US railroads extended over more than 30,000 miles (48,000km). The lines covered most of the eastern states and much of the Midwest, so both troops and matériel could be carried rapidly around the country. Both sides understood the importance of the railroads, but the North made better use of them. Even before the war started, Lincoln ensured that the key railroads were taken under government control and that military traffic was given priority. The outcome of the first major land battle of the war—at Bull Run, Virginia, a small river just 20 miles (32km) south of Washington, DC—was determined by the clever use of the railroads by the Confederates. The battle started as an attempt by the Federal Army to bring the war to a rapid close by capturing Richmond, the Confederate capital. The Federal Army attacked the enemy alongside Bull Run, and initially gained the upper hand. However, they found themselves defeated by a counter-attack, made possible by the quick arrival of Confederate reinforcements via rail from the Shenandoah Valley in the west. This was an important lesson for both sides, and from then on most of the war's major battles took place at or near railroad junctions or stations.

The Federal Army launched the Peninsular Campaign—another effort to capture Richmond—in March the following year, and brought "the war's wizard of railroading" into action. This was Herman Haupt, a brilliant engineer whose background made him the ideal man to harness the railroads for wartime use: he had graduated from West Point, the US Army training college, but then became a professor of mathematics and engineering, and had been appointed superintendent of the Pennsylvania Railroad, one of America's most important lines.

Haupt established two main principles regarding the use of railroads in war. Firstly, the military should not interfere with the operation of services, since it was vital to keep to reliable railroad schedules. Secondly, it was crucial to ensure that empty freight cars were returned to their place of origin and not used as warehouses (or even offices by senior personnel), because running out of freight cars in wartime could mean the difference between success or failure in battle.

Haupt's first task in the Peninsular Campaign was to repair the Richmond, Fredericksburg, and Potomac Railroad, a 15-mile (24-km) vital artery that connected the two capitals, Richmond, Virginia, and Washington, DC. The Confederates had wrecked the line in order to damage the Federal Army's capability, and they had done a particularly

THE WESTERN AND ATLANTIC RAILROAD DELIVERED AN ARMY OF

100,000

MEN TO GENERAL SHERMAN'S FORCES AT ATLANTA

"Haupt has built a bridge… and there is nothing in it but cornstalks and beanpoles"

ABRAHAM LINCOLN, MAY 28, 1862

thorough job, twisting the rails so that they could not be used again and burning down bridges. Several miles of track had been put out of commission, but in response, Haupt performed what seemed to be a miracle. He rebuilt the railroad within two weeks, making it possible for a full complement of up to 20 trains per day to run on the line. His greatest achievement was to erect a 400-ft (120-m) trestle bridge high over the Potomac Creek in just nine days, even though he had only unskilled workers and poor, unseasoned wood at his disposal. This spectacular achievement was lauded by Lincoln when he visited the site: "I have seen the most remarkable structure that human eyes ever rested upon." Perhaps unsurprisingly, however, the president only viewed the bridge from the embankment and did not venture over it.

Haupt's talents included devising methods to destroy the enemy's railroads, a task that proved to be just as important as repairing existing lines and building new ones. Thanks to Haupt's principles, troop movements by rail were carried out without any major mishaps. The greatest movement of the war was when 23,000 men were needed to defend Tennessee after the defeat of the Federal Army at the Battle of Chickamauga in Georgia. The defeated army had retreated to Chattanooga, a rail hub in neighboring Tennessee, and needed reinforcements. In an extraordinary operation, involving seven railroads and two ferry trips, the reserve troops traveled 1,200 miles (1,950km) in just two weeks to relieve the siege of Chattanooga. The town then became a crucial stage for the Federal Army's invasion of the South, which ended the war. That final push was also dependent on the railroads. When General Sherman left Chattanooga on his march to Atlanta, which signaled the end of the war, he relied on the railroad for supplies. He wrote after the war, with military precision:

That single stem of railroad [The Western and Atlantic] supplied an army of 100,000 men and 32,000 horses for the period of 196 days from May 1 to November 19 1864. To have delivered that amount of

THE GENERAL
Buster Keaton heroically lifts a railroad tie out
of the path of his engine in hot pursuit of hijacked
locomotive *The General* in the film of the same name.

forage and food by ordinary wagons would have required 36,800 wagons,
of six mules each... a simple impossibility in such roads as existed
in that region.

A famous episode of the war—immortalized in *The General*, a 1926 Buster
Keaton film occurred on the railroads. A group of 21 Federal soldiers
led by James Andrews penetrated enemy lines at Marietta, Georgia, and
stole a train hauled by a locomotive called *The General*, with the intention
of wrecking the Western and Atlantic Railroad. A Southern conductor,
William Fuller, furious at the hijacking of his train, pursued the hijackers,
first on foot, and then on a gandy dancers' handcart (a device used for
track maintenance). He eventually commandeered a locomotive and,
evading the obstacles placed on the line by Andrews, caught up with *The
General* when it ran out of fuel. The raiders fled into the countryside, but
seven, including Andrews, were caught and hanged, while the rest
escaped to the North. This, however, was a side-show. The lesson of the
war was clear. Railroads were now invaluable military assets, and so
they remained for 100 years or more.

Signals in the Steam Age

In the earliest days of the railroads, trains ran up and down single tracks and signaling was not a necessity. As rail traffic and speeds increased, however, train safety grew to be a major concern and signaling became essential to prevent collisions. Hand and arm signals were soon replaced by flags and lanterns, and in 1832 the first elevated wayside signaling was introduced. By the 1860s, mechanical signals were in general use, but no single system was agreed upon. Semaphore signals were widely adopted in Britain, but were not standardized until 1923, while ball signals were common in the US. Color light signals came into use from the 1950s.

HAND AND ARM SIGNALS
The earliest form of signaling—hand and arm signs—was still in use in the 1930s. Here a brakeman for the Southern Pacific Railroad uses hand signals to hitch cars on a freight train in 1937.

Signaling tokens

A chief safety mechanism in train signaling outside the US is the block system, which allows only one train to enter each "block" of a railroad at a time. In the 19th century, tokens provided evidence that a block was free. In the original "staff and ticket" system, the tower operator gave the locomotive engineer a token or "staff" to allow entry to a block. At the other end, the staff was given up, allowing a train to proceed in the opposite direction. If a second train followed the first along the same "block," both carried written permission or a "ticket." Later systems were operated by means of tokens inserted in a trackside machine.

BALL TOKENS, INDIA

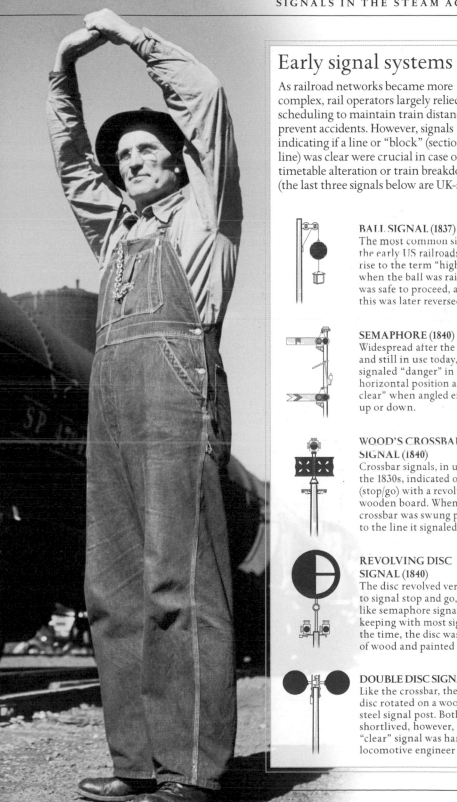

Early signal systems

As railroad networks became more complex, rail operators largely relied on scheduling to maintain train distances and prevent accidents. However, signals indicating if a line or "block" (section of the line) was clear were crucial in case of timetable alteration or train breakdown (the last three signals below are UK-specific).

BALL SIGNAL (1837)
The most common signal on the early US railroads, it gave rise to the term "highball:" when the ball was raised, it was safe to proceed, although this was later reversed.

SEMAPHORE (1840)
Widespread after the 1850s, and still in use today, it signaled "danger" in the horizontal position and "all clear" when angled either up or down.

WOOD'S CROSSBAR SIGNAL (1840)
Crossbar signals, in use from the 1830s, indicated on/off (stop/go) with a revolving wooden board. When the crossbar was swung parallel to the line it signaled clear.

REVOLVING DISC SIGNAL (1840)
The disc revolved vertically to signal stop and go, much like semaphore signals. In keeping with most signals of the time, the disc was made of wood and painted red.

DOUBLE DISC SIGNAL (1846)
Like the crossbar, the double disc rotated on a wooden or steel signal post. Both were shortlived, however, as the "clear" signal was hard for locomotive engineer to see.

Heroic Failures

THE RAILROADS AS WE KNOW THEM were not the only form of rail transportation that was considered in the latter half of the 19th century. Indeed, before the current system became standard, tracks, cars, and engines were combined in all kinds of ways that may seem eccentric today. Some of these ideas might have been successful had they received the proper attention, but others were patently ill-considered and bound to fail.

Perhaps the most spectacular failure was the "atmospheric railway"—a brainchild of British engineer Isambard Kingdom Brunel, the otherwise unimpeachable builder of the Great Western Railway, the SS *Great Britain* (the first propeller-driven iron ship), and the Royal Albert Bridge. His contention was that locomotives were uneconomical because they had to haul themselves as well as their trains (which made uphill travel even harder), and his solution was to remove the engine altogether, creating a train that was propelled by a series of stationary steam engines located along

BRUNEL'S FOLLY
A section of Brunel's atmospheric railroad lies reconstructed in the town of Didcot, Oxfordshire. The vacuum tube in the center contained a piston that hauled the trains.

the track. At the time, locomotives were far from popular—people were afraid of suffocating in tunnels and of sparks setting fire to farmland—so alternatives were welcomed, and in 1844 an Act of Parliament sanctioned Brunel's idea. It was to be tested on the South Devon Railway, a broad-gauge line that Brunel would build from Exeter to Plymouth.

The atmospheric system worked by means of a tube that ran down the center of the track. This functioned like the cylinder of a steam engine (see pp..30–31), having a piston inside it that was connected to the first car of the train (the "piston car"). The connecting arm of the piston traveled through a slit in the top of the tube, which was kept airtight with a long strip of leather and metal components that opened and closed as the piston passed through it. A steam engine at the side of the track created a vacuum inside the tube, which forced the piston forward, hauling the train along with it. The train then traveled some 3 miles (4.8km) until it reached a second engine, which created a new vacuum and propelled the train to the next engine, and so on. A total of 11 engine houses were completed for the line, which took it as far as Newton, about 26 miles (42km) short of the planned terminus at Plymouth—the final series of engine houses was never built.

The railroad opened in September 1847 on the 11-mile (18-km) stretch of line between Exeter and Teignmouth, and at first it seemed to work well. According to Brunel's biographer:

> the new method of traction was universally approved of. The motion of the train, relieved of the impulsive action of the locomotive, as singularly smooth and agreeable; and passengers were freed from the annoyance of coke dust and the sulfurous smell from the engine chimney.

It was also fast, with a top speed of 68mph (109kph) and an average of around half that, which were both remarkable speeds for trains of the period.

There were, however, difficulties from the outset. Setting off from a station proved to be a problem, as the train often needed a helping tow from horses or an extra engine attached to a tow rope. The system was also inflexible, as the pipework prevented the trains from being routed from one track to another. The biggest problem, however, was maintaining the vacuum itself. The leather flaps that sealed the pipe failed to be airtight, and there were rumors that they were gnawed by rats. Likewise the metallic parts of the seal were corroded by salt from sea-spray. Consequently, the atmospheric system was abandoned after a mere eight months, and the equipment replaced by conventional steam locomotives.

It was an expensive failure. The shareholders of the South Devon Railway were almost £400,000 (in modern terms, $55 million) poorer as a result, which was a huge sum at the time. Much of the money had been spent on the elaborate engine houses, which Brunel had constructed in an Italianate style, their large chimneys disguised as campaniles. Each house cost several times the price of a conventional locomotive, and they were so elegant that one, at Starcross, was later used as a chapel. Also, installing atmospheric traction had been nine times more expensive than the original estimate, and the static engines burned far more coal than expected, costing twice that of conventional traction.

Brunel's failure was particularly painful since it happened in public, on a commercial railroad, but there were plenty of other disasters that happened in relative privacy. In 1824, for example, British inventor W.F. Snowden designed a train that eschewed the use of steam power altogether. It had a single line of wheels that ran in a U-shaped rail flanked by a pair of flat rails that kept the cars upright. To propel the train, "industrious laborers" in the lead car literally cranked a wheel that was connected to a gear that engaged with the toothed edge of one of the rails—thus providing human traction, and requiring superhuman stamina on the part of the laborers. Not surprisingly, the idea was never taken up, despite a pamphlet published in 1834 extolling its virtues. However, Snowden's system did highlight a problem that is common to all railroads—that wheels

"He had so much faith in his being able to improve it that he shut his eyes to the consequences of failure"

DANIEL GOOCH, BRUNEL'S COLLEAGUE, ON THE ATMOSPHERIC RAILWAY

slip in damp conditions and on inclines—a problem Snowden solved by keeping the gear locked in the track. A related idea was to have a rack and pinion (see pp.108–109) in the center of the railway to aid traction, and such devices are still used on mountain lines today.

Rope or cable railroads were another attempt at solving the problem of how to haul trains. Again, as with atmospheric railroads, the idea was to have stationary engines placed along the track, but this time for hauling cables attached to the trains. On the Canterbury and Whitstable line in Britain, which opened in May 1830, trains were hauled by rope for 4 miles (6.5km) out of Canterbury and then by locomotive for the remaining two miles. Several other lines used cables to deal with inclines. Nearly two miles of the Düsseldorf–Elberfield line in Germany was cable-hauled, as was part of the Brussels–Liège line in Belgium, and the Denniston Incline on the South Island of New Zealand. The Liverpool and Manchester (see pp.25–29) had cables for the first part of the line out of Liverpool station, as did the London and Birmingham for the incline between Euston station and Camden Town. One of London's first railroads, the London and Blackwall, was cable-drawn for its entire 3½-mile (6-km) length. Cables were complicated to operate, however, and as locomotives became more powerful, most of these systems were phased out (although the German one remained in operation until 1927). A cable system was even proposed for the first deep subway line built on the London Underground—the City and South London line, completed in 1890—but given its length of nearly 5 miles (8km), planners decided to use the new technology of electric power instead.

Monorails were another great hope of the railroad pioneers, and although a few were constructed—and indeed some still operate—they have never overcome the basic problems of being expensive to build, and being inflexible due to the structural requirements of their rails. The first patent for a vehicle designed to run on a single rail was granted in November 1821 to British civil engineer Henry Robinson Palmer, who described it as "a single line of rail, supported at such height from the ground as to allow the center of gravity of the carriages to be below the upper surface of the rail." The vehicles straddled the rail, rather like pannier baskets on a mule, and were horse-drawn. The idea was to make it easier to transport goods across worksites, and the first monorail was built in the Deptford Dockyard, London, in 1824. The following year,

another line, the horse-drawn Cheshunt Railway, was built at a brick factory in London. At its opening it carried passengers—an historic moment, as it predated the world's first passenger railroad, the Stockton and Darlington (see pp.24–25), by three months. Monorails have since resurfaced from time to time, but never with much success. A steam-driven monorail was first demonstrated at the United States Centennial Exposition in Philadelphia in 1876, and although a couple of versions were built, neither lasted very long. The oldest monorail in the world is the Wuppertal Suspension Railway in Germany, which opened in 1901 and still operates today, carrying 25 million passengers a year. In the later 20th century, a number of monorails were built in urban areas in Asia, including Tokyo in Japan and Kuala Lumpur in Malaysia.

ROPE AND STEAM
Propelled both by cables powered by stationary steam engines, and moving steam locomotives, the Canterbury and Whitstable Railway was known locally as the "Crab and Winkle" line.

Fanciful ideas for new transportation systems continued into the 20th century. Perhaps the strangest of all was the balloon railroad constructed near Salzburg in Austria in the early 1900s. This consisted of a large balloon tethered to a slide running on a single rail up a mountainside. The hydrogen balloon hovered some 33ft (10m) above the car, which could carry up to ten passengers. Once loaded, the balloon was freed, pulling up the car beneath it. To descend, the car's tanks were filled with water. Its inventor, Herr Balderauer, believed that his system would replace the costlier funiculars that were being built across the Alps (see pp.102–107), but perhaps unsurprisingly, it failed to attract investors.

AN UNUSUAL SUCCESS
The German suspended railroad, the Wuppertal Suspension Railway, photographed in 1912. It has run for over a century and continues to operate today.

India: Dalhousie's
Colonial Imperative

INDIA WAS THE JEWEL IN BRITAIN'S CROWN during the
19th century, and its colonial rulers were eager to impose their
dominion over the subcontinent. The railroads were vital to this process
and were brought to India by the British East India Company – effectively
a commercial arm of the British government. The first Indian railroad,
which ran from the company settlement of Bombay, was commercial in
aim, prompted by events on the other side of the world: the American
cotton harvest had failed in 1846, and this spurred the cloth manufacturers
of Manchester to use India as an alternative source of cotton. However,
in order to ensure the steady supply needed to keep their factories
running, transportation to the Bombay port needed improving, and so
the cotton magnates pressed the British government to build a railroad.

Due to bureaucratic delays—it took months to receive a reply to a
letter sent from India to Britain—and indecision on the part of the
British rulers, work on the 21-mile (34-km) line between Bombay and
Thane, or Tana, did not start until 1850. It was an experiment to see if it
was possible to build railroads in the harsh climate of the Indian
subcontinent. The Thane line was not easy to construct. Now part of
Bombay's busy suburban network, it originally went through difficult
territory, including a hill that had to be cut through and a marsh.
Nevertheless, it was completed in three years.

The opening of the line in April 1853 was a momentous affair, not
least because it was the first railroad in Asia. Unlike northwest
England, where railroads had originated, India was still a rural,
unindustrialized nation, and the sight of powerful locomotives
belching fire and steam impressed and frightened the local populace
in equal measure. People lined the tracks by the millions to watch the
14-passenger car train bearing a rich assortment of VIPs on its
inaugural journey, and many spilled onto the tracks, slowing the
progress of the train, which nevertheless managed the journey at an
average speed of 20mph (32kph)—a very creditable effort.

Another, more ambitious, project started simultaneously in Bengal,
in the northeast of India. This was a 121-mile (195-km) line stretching
from Howrah, on the western side of the Hooghly river, via nearby

Calcutta to the small town of Raniganj in the coalfields of Burdwan, from which it had previously taken two seasons to cart the coal down to a river for transportation to the rest of the country. Work on the line started in 1851, but a couple of shipping mishaps delayed progress. First, the passenger cars intended for the line were lost when the ship carrying them sank at Sandheads on the Bengal coast, and then, astonishingly, the locomotives for the line were sent to Australia instead of Calcutta as a result of one of the most expensive clerical errors in history. The locomotives eventually arrived in Calcutta—a year late—but, despite these setbacks, the line, nearly six times longer than the Bombay–Thane railroad, opened in February 1855.

The success of these two lines led to a rapid expansion of the railroad system. Unlike in Britain, where the process was unplanned and haphazard, the development of the rail network in India followed a clear plan laid out by the colonial administration. In 1853, Lord Dalhousie, the governor-general, set out a program for the development of India's trunk lines in a 216-page handwritten memorandum. Dalhousie was a hard-working, capable administrator who later claimed to have given India all the necessary "engines of social improvement... Railways, uniform postage, and the Electric Telegraph." His "minute," as he called the memorandum, set out the justifications and guidelines for the railroads of India, whose territory then included what is now Bangladesh and Pakistan.

Dalhousie emphasized the economic benefits that the railroads would bring. They would increase trade between India and the mother country—Britain importing cotton and India receiving manufactured goods in return. Railroads would encourage enterprise, increase production, facilitate the

LORD DALHOUSIE
Governor-General of India from 1847 to 1856, Lord Dalhousie set out a visionary plan of railroad building in India that aimed to rival the wonders of the ancient world.

THE EAST INDIAN RAILWAY COMPANY WAS ESTABLISHED IN 1845 WITH A CAPITAL OF

£4,000,000

25,000

LABORERS DIED BUILDING THE WESTERN GHATS RAILROAD

discovery of natural resources such as coal and minerals, and encourage overall economic development as they were doing in Europe and the United States, which were now both three decades into the railroad age. However, the core of Dalhousie's argument was not commercial but political. The British administrators and soldiers who ruled the country were thinly spread, and railroads would enable them to travel rapidly to maintain control—an advantage that was worth almost any price.

For his part, Dalhousie was nothing if not a visionary. In his minute, he wrote enthusiastically:

the complete permeation of these climes of the sun by a magnificent system of railway communication would present a series of public movements vastly surpassing in real grandeur the aqueducts of Rome, the pyramids of Egypt, the Great Wall of China, the temples, palaces and mausoleums of the great Moghul monuments.

An exaggeration, perhaps, but it gives an idea of the enormity of the plan. Dalhousie's blueprint was given urgency by a mutiny of the army's Indian soldiers that began in May 1857 and threatened colonial rule. The mutiny—which began with soldiers refusing to bite on musket cartridges covered with tallow made from either beef fat (offensive to Hindus) or pork fat (offensive to Muslims)—had broken out in remote parts of the country. The military realized that rather than bringing vast numbers of British troops over to manage the Indian regiments, it was cheaper to build railroad lines to ensure that troops could be despatched at speed to deal with insurrections wherever they happened. The mutiny halted work on railroad projects across India, but work recommenced at a faster pace when peace was restored in June 1858.

The railroads, in other words, were a nakedly colonial project that paid scant regard to the needs of the native population. The British decided on the location of the lines and when they would be built. They were designed to serve British interests and would later be seen by nationalists such as Mahatma Gandhi as agents of imperialism. The British government accepted Dalhousie's plan and the railroads were constructed to the pattern he had devised. Major trunk routes radiated from the port cities and centers of colonial administration, such as Calcutta, Bombay, and Madras, and further lines connected other large towns and cities. It was, as one writer suggests, a period of great "romance and excitement," but it was also a time of hardship, particularly for innumerable Indian workers, thousands of whom died constructing the railroads. The country's diverse and rugged landscape also posed formidable challenges. Bridges, for example, had to withstand far greater pressures than those of Europe, because the monsoon rains made rivers swell to enormous, and destructive, proportions.

COLONIAL RAIL
This plate taken from the *Illustrated London News* shows the arrival of a railroad locomotive in India in 1875. By 1880, India had around 9,000 miles (14,500km) of track.

The Western Ghats, a chain of mountains that run down much of the western side of the subcontinent, proved to be an almost insuperable barrier. Although not immensely high—a mere 8,840ft (2,695m)— they rise sharply and ruggedly beyond the narrow coastal lowlands, presenting what at the time was the most difficult section of railroad terrain in the world. It took engineer James Berkley several years just to survey the route, although the line was only 15 miles (24km) long. When it was finished, the line featured two major inclines (see pp.204– 205)—the Bhore and the Thul Ghats—as well as numerous tunnels. Berkley devised an ingenious way of overcoming the steep grade. Instead of building a continuous track, he carved out a reversing section at a bend near the summit, obviating the need for a stationary engine to haul the trains uphill. It was a cheaper and neater solution than a conventional route, and was imitated on several other mountain lines, notably in Brazil and the Andes. The system involved driving a train along a narrow line toward a precipice, so it required nerves of steel on the part of the engineer, and indeed the passengers.

BHORE GHAT REVERSING INCLINE
The reversing station (see pp.204–205) on the Bhore Ghat incline in the Western Ghats, seen here in 1880, was an ingenious way of negotiating the Indian hills, although it involved a hair-raising reverse toward a precipice.

"If we did not rush about from place to place by railway, much confusion would be obviated"

MAHATMA GANDHI, *THE ESSENTIAL WRITINGS*

To create the track bed for the Ghat line, whole sections of the mountain had to be blasted and workers let down on ropes to drill into the mountain face—a perilous process. On numerous occasions, the ropes failed or slipped, sending workers down into a ravine from which their bodies were never recovered. Illness, however, was the biggest killer. Numerous killer diseases, including typhoid, malaria, smallpox, cholera, and blackwater fever, took a high toll on the overworked and undernourished workers, who died by the tens of thousands. The lives of the Indian workers, or "coolies," as they were known, were considered cheap in Colonial India. As one government report tellingly reads:

> The fine season of eight months [work except in tunnels halted during the monsoon] is favourable for Indian railway operations, but on the other hand, fatal epidemics, such as cholera and fever, often break out and the labourers are generally of such feeble constitution, and so badly provided with shelter and clothing, that they speedily succumb to those diseases and the benefits of the fine weather are, thereby, temporarily lost.

As Anthony Burton, the author of a book on railroads and the British Empire, observes: "the notion that lives—and the inconvenient loss of working time—could be saved by providing proper shelter and decent conditions does not seem to have been considered."

The spread of the railroad in India came at a high cost, but it remained a remarkable achievement, and Dalhousie's plan was adhered to long after his departure and death. His belief that the project would be bigger than the construction of the pyramids was borne out. Within 25 years of the opening of the Thane line, India had an extensive network of trunk lines. By the turn of the century, there was a remarkable 25,000 miles (40,000km) of track—a network that has survived to this day with few closures, and which remains a key part of the infrastructure of a nation in which a highway network was only developed late in the 20th century. The Indian railroads have become synonymous with India itself.

Early Indian Railroads

Railroad construction took place at a rapid pace on the Indian subcontinent from the 1850s onwards. Following Dalhousie's recommendation of a series of arterial "trunk" routes to connect India's major cities, the first lines were built inland from the regional hubs of Bombay on the west coast, Madras in the south, and Calcutta in the east. Smaller regional lines—built in an alarming variety of gauges—fanned out from these main lines, and by the early 20th century India boasted more than 25,000 miles (40,000km) of track. This map shows the main rail network planned by Dalhousie and built in the colonial era.

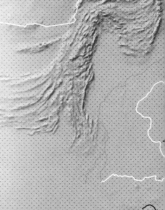

Jamnagar

Rajko

ARABIAN
SEA

KEY

●   Major city

○   City/town

———   Main line

———   National boundary

INDIAN
OCEAN

CROSSING THE RIVER
A steam locomotive is floated across the Yamuna River on a pontoon at Kalpi, northern India, in the late 1800s. The river was later bridged, but before and during this contruction, workers had to use improvised measures such as this to transport locomotives over the river.

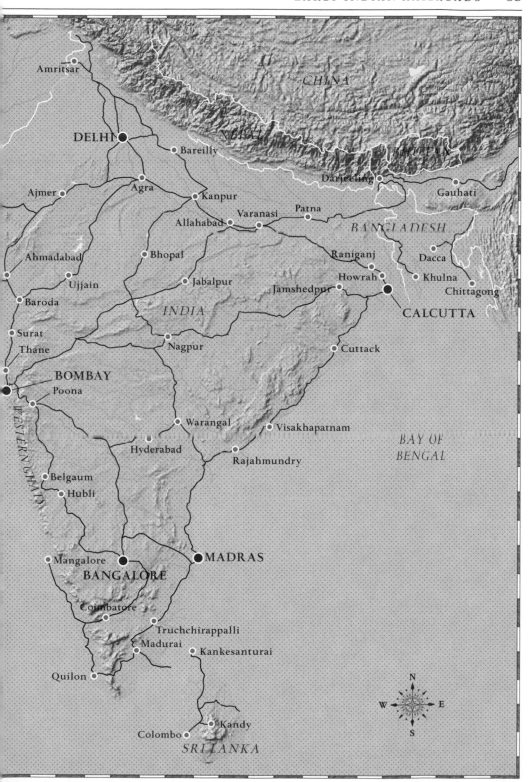

Amritsar

CHINA

DELHI

Bareilly

NEPAL

BHUTAN

Ajmer

Agra

Kanpur

Darjeeling

Gauhati

Allahabad

Varanasi

Patna

BANGLADESH

Ahmadabad

Bhopal

Raniganj

Dacca

Ujjain

Jabalpur

Jamshedpur

Howrah

Khulna

Baroda

INDIA

Chittagong

Surat

Nagpur

Cuttack

CALCUTTA

Thane

BOMBAY

Poona

WESTERN GHATS

Warangal

Visakhapatnam

BAY OF
BENGAL

Hyderabad

Rajahmundry

Belgaum

Hubli

Mangalore

MADRAS

BANGALORE

Coimbatore

Truchchirappalli

Madurai

Kankesanturai

Quilon

Colombo

Kandy

SRI LANKA

N
W   E
S

The Navvies: Digging, Drinking, and Fighting

THE MEN WHO BUILT THE RAILROADS were a tough bunch— and they needed to be, as they had an arduous job, carried out in remote areas and often in harsh conditions. They were also at the cutting edge of technology, working in a new industry that had developed its own machinery and working methods. It was a learning process for all concerned, from the contractors and engineers to the men who laid out the embankments and dug the tunnels. Astonishingly, the tracks used by modern, high-speed trains would be quite recognizable to the engineers who built the first lines. After a few early experiments with granite ties and wooden rails, the basic design was adopted almost everywhere in the world—iron (later steel) rails laid on wooden ties, resting on stone aggregate or ballast (see pp.90–91).

For much of the 19th century, laying the tracks was very labor-intensive. A surveyor drew an approximate line on a map after walking the site, then thousands of workers were hired by a contractor. The workers were called "navvies" because they were thought to have the same skills as the navigators who built the canal system a generation earlier. These navvies were proud of their name, but by no means all the workers on the railroads qualified for it. According to Terry Coleman, author of *The Railway Navvies*, the key book on the history of the navvies, they "must never be confused with the rabble of steady, common laborers whom they out-worked, out-drank, out-rioted and despised." The laborers came and went, many returning to the farms at harvest time. If they stayed, however, it took a year for laborers to qualify as navvies, who were considered an elite class of worker. To be navvies, they had to work on all the hard tasks, such as tunneling, excavating, and blasting, and not simply shoveling earth; they had to live with the other navvies in camps and follow the railroad as worksites moved along; and they had to match the eating and drinking habits of their fellows— consuming nearly 2lb (1kg) of beef and 9½ pints (4.5l) of beer a day. The navvies came from all over Britain and adopted their own particular dress code. According to Coleman, they favored "moleskin trousers, double-canvas shirts, velveteen square-tailed coats, hobnail

NAVVIES USING A BRICKMAKING MACHINE Although their reputation as volatile, hard-drinking fighters was often deserved, the navvies also carried out long hours of hard labor, and had to put up with poor working and living conditions.

boots, gaudy handkerchiefs and white felt hats with the brims turned up." It was not really suitable garb for such hard, manual work, but it demonstrated style. They were known by nicknames that ranged from the eclectic "Bellerophon" or "Fisherman" to the more common "Gipsy Joe" or "Fighting Jack."

To some extent, the skills needed to build the railroads were tried and tested. There were similarities with building canals and digging out mines, but in terms of scale, perhaps only cathedral-building compared—although cathedrals took centuries to complete, while railroads took only a few years. The scale and extent of the earthworks alone was unprecedented. The most visible features of the navvies' trade were bridges and tunnels, but the vast majority of their work consisted of moving enormous quantities of earth. Railroads require relatively straight routes and gentle grades, so the land has to be adapted before the track can be laid. Peter Lecount, an assistant engineer on the London & Birmingham, calculated that building it involved lifting 25,000 million cu feet (708 million cu m) of earth—a greater task than the building of the Great Pyramid at Giza. Occasionally, new techniques were deployed, such as laying embankments through marshy land, or crossing large rivers with bridges. Worksites commonly had hundreds of men attacking the earth with their primitive tools and hauling the dirt away in wheelbarrows, or by horse and cart on the flatter sections. As railroad historian R.S. Joby recounts:

rough-looking gangers drive their butty gangs to ever greater feats of earth-moving by threats, promises, and, at times, a well-timed kick for the pay of the gang depends on the efforts of the team as a whole.

A great deal of gunpowder was used, the fuse-setters and navvies crouching behind any available cover to shield themselves before returning to prepare the next blast. Safety precautions were minimal and risk-taking was considered manly. Not surprisingly, navvies often died young. In the Kilsby Tunnel on the London & Birmingham, three men were killed as they tried to jump, one after the other, over the mouth of a shaft in a game of "follow my leader." On the Great Western, a man was told to stop cutting under a large overhang of earth, but he ignored the warnings and was buried alive within

INTO THE DARK
Tunneling under Blackfriars Bridge, London, in the 1860s. In a life filled with risk and tough conditions, working in tunnels was the epitome of both for the navvies. Accidents were frequent.

minutes. Those who survived the numerous accidents—ranging from unexpectedly large explosions to simple falls and collisions— were soon worn out by the hard work and the excessive lifestyle. The forty-year-olds looked fifty, and the rare sixty-year-olds looked eighty. The sheer numbers of workers involved in railroad-building was also colossal—far greater than in any previous industry. In spring 1847, for example, 169,838 men were working on the railroads in England and Wales, out of a total population of 16 million. Initially, the companies employing these men were small, local concerns, but large contractors soon emerged, employing thousands of navvies. These contractors were powerful men such as Samuel Peto and Thomas Brassey, who built railroads all over the world.

Not surprisingly, navvies were rarely welcome in the towns where they were working. They caused a great deal of disruption, and shopkeepers took advantage of the surge in demand to raise food prices. As accommodation was scarce in the remote places where lines were being built, the navvies slept in huts in filthy conditions, sometimes sharing their quarters with pigs and attracting vermin. They often shared beds—one man sleeping while the other worked— and they were accompanied by a retinue of "many women but few wives," as one writer put it. As a result, disease was rife. Riots were not uncommon, particularly on payday, or, as sometimes happened, when the navvies had *not* been paid. In 1866, the village of Wiveliscombe in Somerset, England, was terrorized by 70 navvies demanding beer and bread after the local railroad company went bust. A local resident of a Devon village described the chaos when the navvies found themselves without work after the railroad was finished:

> More than a hundred discharged on Monday, and a pretty row there was: drunk altogether and fighting altogether, except one couple fought in the meadow for an hour… the same night the villains stole all poor old xxx's fowls [and] there is not an egg to be got hereabouts.

Similar scenes were played out in other countries. The navvies may have been wild, but they got the job done, so much so that British men found work on many European railroads. In 1843, Thomas Brassey, a contractor, was commissioned to build the line between Rouen and Le Havre in northern France, and a local newspaperman sent to observe construction was greatly impressed:

I think as fine a spectacle as any man could witness, who is accustomed to look at work, is to see a cutting in full operation with about twenty wagons being filled, every man at his post, and every man with his shirt open, working in the heat of the day, the gangers looking about, and everything going like clockwork. Such an exhibition of physical power attracted many French gentlemen who came on to the cuttings at Paris and Rouen, and looking at these English [actually many were Scottish and Irish] gentlemen with astonishment said 'Mon Dieu! Les Anglais, comme ils travaillent!' [My God! The English, how they work!]… It was a fine sight to see the Englishmen that were there, with their muscular arms and hands hairy and brown.

There were often labor shortages in the US, and men had to be brought in from other countries to build the railroads. The construction of the Erie Railroad in the late 1830s and early 1840s coincided with a large influx of men from Ireland who were fleeing the famine at home and were eager to work on the railroad. Unfortunately, the men were from two different parts of Ireland— Fardown and Cork—and the former took against the latter in a dispute over lower wages. In the ensuing conflict, the Fardowners set upon the Corkonians in a series of battles that lasted several days. This culminated with the Fardowners cutting down the rickety structures in which the Corkonians lived, bringing the roofs down on top of them. It is a wonder that they found enough time to build the railroad at all, but construction was unaffected.

In the 1860s, the Central Pacific company was building the line eastward from California on the first transcontinental (see pp.120–27). However, the construction was desperately undermanned, due to both a lack of immigrants and the competition from the lucrative mining industry. One of the line's promoters named Charles Crocker hit upon the idea of taking on Chinese workers, but had to overcome resistance from his worksite managers who thought that these "tiny rice-eaters" could not handle such work. They were proved wrong, however: the Chinese were extremely good laborers and

TYPICAL AMOUNT OF EARTH MOVED BY ONE NAVVY IN A DAY

4,400 LB
(2,000 KG)

RAILROADS ON THE STEPPES
Convicts build the Ussuri section of the
Trans-Siberian Railway in the far east of
Russia in the early 1900s.

accepted lower pay than their white counterparts, prompting Crocker
to organize the recruitment of thousands of men from China. Chinese
laborers also worked on the legendary South American railroads,
accompanied by the fearsome local workforce, the *rotos* (see pp.202–203).

In Russia, the construction of the Trans-Siberian Railway in the
1890s (see pp.180–87) also meant that a lot of men had to be brought
in from afar. The line, which is still the longest in the world, required
huge numbers of workers and had 80,000 enrolled at its peak. The
local people, mostly tribesmen, were unwilling or unable to work,
and as the line progressed further east into the almost deserted steppe
the shortage of labor was acute. Convicts were drafted in, with 13,500
prisoners and exiles working on the line at the peak of its construction.
Laborers also had to be brought in not only from European Russia,
but from as far afield as Turkey, Persia, and Italy.

This conscription of workers continued throughout the 19th century,
after which mechanization reduced the need to mobilize such vast labor
forces. Until then, however, the railroads were the work of strong men
wielding primitive tools. Their legacy can still be seen, not only in
railroads that have since been modernized, but in the embankments
and cuttings that remain where lines have been closed. All over the
world, the landscape was transformed by their herculean labors.

The Track Structure

The track structure is comprised of track components (e.g., rail, ties, fasteners, and joint bars) and trackbed structural components (ballast, sub-ballast, and subgrade). In the early days of railroad construction, temporary track was laid first so materials could be transported quickly to the site. After the substructure was complete, the permanent track was laid. The substructure of a track is called the formation. Since a consistent grade is required in order for trains to run smoothly, the ground is first prepared to form the subgrade. The subgrade might also be covered by a layer of sand or stone called a blanket before overlaying it with ballast. The gauge (distance apart) and alignment of the rails are monitored to ensure that they remain constant throughout straight sections and curves in the line.

LAYING THE TRACKS
Workers follow a track-renewal train as it places rails on newly laid ties. Before the mid-1900s, workmen carried out the process by hand. These bullhead rails had the same profile on the head (top) and foot (underside), allowing them to be inverted and reused when the head wore out.

Track materials

Wooden rails were used for the pony-drawn wagonways of the 17th century, but a more lasting material was required to support 19th century steam engines. The cast-iron rails of the first railroads were succeeded by sturdier wrought-iron rails in the 1820s, while steel—which was stronger still—came into use in the 1850s. Crushed-stone ballast is still the most common foundation material, but concrete slabs, which are more stable and durable, are increasingly used.

TRACK STRUCTURE
Most railroad tracks consist of flat-bottom steel rails fixed to wooden or concrete ties. The layer of ballast beneath has the benefit of reducing noise from rail traffic, but requires maintenance due to displacement from the weight of passing trains.

Rails

Tie

Track

Shoulder

cess   BALLAST/SUB-BALLAST

BLANKET (SAND, OPTIONAL)

Track foundation

Formation

SUBGRADE (LOCAL MATERIALS SUCH AS TOPSOIL)

GROUND LEVEL

Track gauges

The gauge of a track is the horizontal distance between the inside faces of the two rails, and determines the axle width of trains that can run upon it. A range of gauges was used when the railroads were built in the 19th century, but standardization became necessary as individual lines were connected to form national and international networks. As such, standard or international gauge (4ft 8½in or 1,435mm) is used for around 60 percent of the world's railroads.

COMMON GAUGES

- 4ft 8½in (1,435mm)
- 4ft 11⅘in (1,520mm)
- 3ft 3⅜in (1,000mm)
- 3ft 6in (1,067mm)
- 5ft 6in (1,676mm)
- 5ft 5⅔in (1,668mm)
- 5ft 3in (1,600mm)
- 3ft 1⅖in (950mm)
- Other gauges

Cuban Sugar Railroads

FEW NATIONS REMAINED UNTOUCHED by the railroad boom of the 19th century, and many were affected in remarkably different ways. Cuba was the first country in Latin America to build railroads, but their purpose was very different from those of the US and Europe. Their chief role was to transport sugar—a crop that was grown in vast plantations to satisfy the sweet teeth of the developed world—and so they did little to help the general population. Cuba had become a sugar economy somewhat by accident, after an increase in the price of sugar in the late 18th century made its cultivation highly profitable. Relying heavily on slave labor, the industry grew rapidly in the early 19th century, but transporting the cane to the mills was expensive because of the island's poor roads. Indeed, in the rainy season, from May to November, the tracks became muddy and rivers flooded, making movement almost impossible. What was needed was a railroad to shift the cane swiftly and cheaply—from the plantations to the mills for processing, and then on to the coast where it could be shipped

CUBAN SUGAR RAILROAD
Constructing the first Cuban railroad, between Havana and Bejucal, involved a massive workforce of slaves from the plantations and Irish immigrant laborers.

150 YEARS OF CUBAN RAIL
These commemorative stamps
were isued in 1987 to mark the
150th anniversary of the opening
of the first Cuban railroad line,
one of the earliest in the world.

abroad. Such a line was duly built, and the crop that was brought to
Cuba by Christopher Columbus in 1493 found its way back to Europe in
unprecedented amounts, making fortunes for Cuba's sugar barons.

Throughout the 19th century, Cuba was a Spanish colony, but its
railroad opened long before that of its imperial master. Indeed, although
Cuba was undeveloped and wretchedly poor at the time, it was one of
the first countries in the world to build a railroad—by the time its first
line opened in 1837, only six other countries had railroads. In 1830, just
as the Liverpool and Manchester line was being completed in Britain
(see pp.25–29), leading plantation owners in Cuba set up a commission
to consider building a railroad network. In 1834, once a route had
been established and sufficient funds raised, work started on a 46-
mile (74-km) line between the capital, Havana, on the coast, and
Güines, inland on the Mayabeque River.

It was a grand and sophisticated operation. The first part of the
16-mile (26-km) route, from Havana to Bejucal, climbed to 320ft
(98m) above sea level, a very steep grade for a railroad at the time.
The line also had several bridges, the longest of which, across the
Almendares River, needed 200 supporting pillars. Moreover, unlike
many other early lines, the Cuban railroad was designed to have two
tracks right from the start. To supplement the thousands of slaves
owned by the railroad company, workers were brought in from
abroad to help construct the line. These were largely Irish immigrants
who had only recently reached the US, and men from the Canary
Islands, which was also under Spanish rule. The new arrivals did not
thrive. The Irish in particular suffered, being unused to the tropical
climate, which was at its worst during the rainy season. Badly fed
and poorly sheltered, large numbers of workers succumbed to the
effects of tropical diseases. Many of them also turned out to be
drunks and quickly found themselves in the filthy colonial jails,

where they soon perished. The survivors found that when their contracts ended the railroad companies did not fulfil their promise of returning them to the US, and they were left to wander the streets of Havana, penniless and destitute. The Canary Islands immigrants fared little better, even though they spoke Spanish. Their employers considered them more likely to try to escape, so they were treated like prisoners on the work sites. Forced to work up to 16 hours per day, many of them died of exhaustion.

The project ran out of money, but new investors stepped in and the first section of the Bejucal line was completed by 1837, just three years after work had begun. The locomotives and engineers were imported from England, and when the section to Güines opened the following year, the railroad boomed. Although it was primarily intended for freight, passengers flocked to the railroad too, generating as much revenue as the sugar in the early years. There were two trains a day in each direction—a 30-car freight train, and a seven-car passenger train.

The success of the line prompted further railroad construction. The first section of a second line—built to bring sugar, molasses, and rum to the port of Cárdenas—was completed in 1840, after which the system expanded rapidly. A decade after the first line opened, the area around Havana was criss-crossed with lines connecting all the neighboring districts, and the sugar industry boomed as a result. In 1846, Havana alone had 169 sugar mills producing more than 40,000 tons of sugar and 45,000 barrels of molasses each year. The railroads serving these mills were highly profitable, but they were geared so specifically toward the sugar industry that they provided little support to other industries, or indeed to passengers. They were built purely to help sugar merchants export their produce, and so were not treated as commercial enterprises. Unlike in Europe and the US, where every railroad junction soon became a bustling town, the railroads did not stimulate urban development in Cuba.

By 1852, nine companies had built a total of 350 miles (565km) of track, although the pace of

APPROXIMATE NUMBER OF AFRICAN SLAVES TRANSPORTED TO CUBA

800,000

THE CUBAN SUGAR RAILROADS

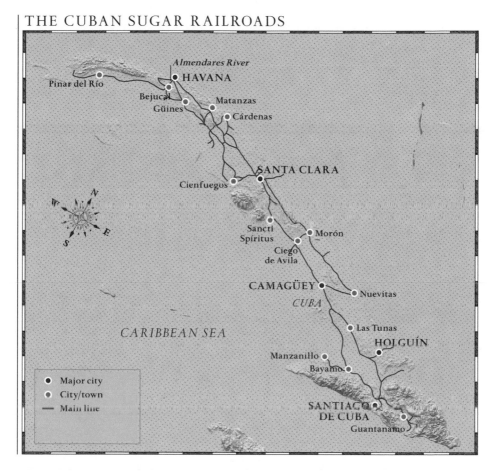

Almendares River
Pinar del Río
HAVANA
Bejucal
Güines
Matanzas
Cárdenas
SANTA CLARA
Cienfuegos
Sancti
Spíritus
Morón
Ciego
de Avila
CAMAGÜEY
Nuevitas
CUBA
Las Tunas
HOLGUÍN
CARIBBEAN SEA
Manzanillo
Bayamo
SANTIAGO
DE CUBA
Guantánamo

- Major city
- City/town
— Main line

growth slowed down due to a slump in world sugar prices and a clampdown on slavery. However, sugar prices recovered, stimulating a second boom in railroad construction. The Crimean War of 1854–56 raised prices even higher—sugar was already a global commodity, and the war diverted British shipping from carrying sugar from Asia, creating a surge in demand. The sugar barons found themselves with enormous amounts of cash, which they used to invest in more and more railroads. Consequently, the amount of track laid reached 800 miles (1,288km) in 1868, providing this impoverished island in the Caribbean with one of the densest railroad networks in the world. It was bettered only by a few major European nations, and in fact had more miles of track per inhabitant than any other country in the world—more, even, than England, the mother of the railroad.

CUBAN SUGAR MILL
An American locomotive waits outside a sugar boiling house in Cuba in 1857. Such locomotives took sugar from the mills to the ports for export.

By the late 19th century, the Cuban railroads had reached a remarkable 5,000 miles (8,000km) of track, half of which were standard-gauge lines designed to shift sugar out of the plantations and sugar products from the mills. The rest were mainly narrow-gauge tracks, often crudely laid, for hauling cane on the plantations themselves. Unfortunately, however, not only did the railroads fail to serve passengers, they failed to stimulate the economy, for they remained dependent on British and US technology: no Cuban supply industries were ever established. If anything, the railroads had a rather damaging effect on the wider economy—they exacerbated the differences between the rich areas, which benefited greatly from their construction, and those that were poor, which became even more neglected. They were also concentrated in the affluent western half of the island. Only a handful of lines were built in the east, where there were few plantations. Crucially, the east–west line connecting the two halves of the island was not built until the 20th century, because it required a government subsidy, which was not forthcoming. As the

BEGGING IN HAVANA
At the height of its success as a sugar economy,
Cuba was awash with foreign laborers, many
of whom never returned home, and ended up
begging on the streets of Havana.

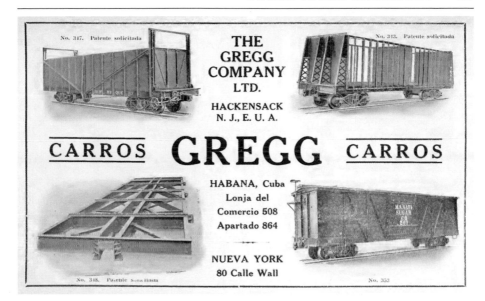

GREGG COMPANY RAILROAD CAR
Like most of Cuba's railroad technology, this freight car, used
for transporting sugar cane, was imported from the US. Cuba
never developed its own train manufacturing industry.

authors of a study of the Cuban railroads argue: "the railroad
development of the first decades lacked the long-term perspective
that would permit the growth of a national grid." In fact, the economic
effect of the railroads was quite perverse. By making the plantation
owners so rich, they helped perpetuate the slave system that might
otherwise have collapsed.

The railroads' dependence on the sugar trade was ultimately their
undoing. They were profitable as long as sugar boomed, because there
was no other form of transportation. In the late 19th century,
however, another collapse in sugar prices rocked Cuba's economy,
leading to the takeover of the railroads by British investors. Few lines
were found to be viable, so much of the network was shut down. By
the early 20th century, the remainder of Cuba's railroads were entirely
owned by British and US companies. These monopolized the western
and eastern networks respectively, but the stagnation of Cuba's sugar
industry drove the railroads into crisis. The spread of the motor car
and improvements in Cuba's roads only deepened the railroads'
problems, which would only be solved when the network was
nationalized in the late 1950s.

The Spread of the Railroads

TAHAO, NO.20
V&T RAILROAD
STEAM, 1875

As train travel increased and the benefits of rail links became evident, it seemed that nothing could stop the spread of the iron road. There was no obstacle—financial, geographic, or social—that could not be overcome by the railroads. Soon mountains were being crossed or tunneled under in territories as far apart as the Austro-Hungarian Empire and India. Rivers were forded and houses demolished to make way for stations in town centers. Even disease-infested jungles were conquered, such as the swamps on the Panamanian isthmus, although only at a terrible human cost. The United States, with its vast western deserts, soon boasted not just one transcontinental railroad, but four—and Canada laid three of its own. Urban transportation, too, was revolutionized. London's Metropolitan Railway, the world's first underground line, opened in 1863, and became the blueprint for many such systems across the world.

Although trains were becoming commonplace, there was little improvement in the quality of services—largely because the majority of travelers had no other means of transportation, so their custom was taken for granted. There were exceptions, of course prestige services such as those developed by George Pullman, who provided not only far better meals, but comfortable overnight sleeper cars—but by and large rail transportation offered few comforts. Nor was it entirely safe. At first, trains were so few and so slow that collisions were unlikely, but as the tracks filled up and trains traveled faster, accidents became inevitable.

The railroad companies soon became the dominant industry of the day. They were larger than any other business, and by their very nature operated across vast areas. Perhaps most enduringly, they liked to demonstrate their importance by building huge stations that became the cathedrals of the age—a source of pride to both the railroad companies and the communities they served.

Crossing the Alps

MANY OF THE FIRST EUROPEAN RAILROADS ran from cities to ports, so that goods could be transferred onto ships. As the network developed, however, it hit a major hurdle in the heart of the continent—the Alps. From the earliest days, the governments promoting the railroads realized that this was an obstacle that would have to be surmounted, but it posed the hardest challenge yet encountered by railroad builders. Engineers had to develop new skills and techniques, excavating tunnels far longer than any previously cut and erecting bridges over deep, inaccessible ravines.

The first railroad to cross the Alps was the Semmering. It was built over the Semmering Pass by the Austrian Empire to connect the imperial capital, Vienna, with the Empire's only seaport, Trieste (now in Italy). A circuitous route via the Hungarian plains had been considered, but the transalpine railroad's main backer, Archduke John of Austria, was determined to find a way over the mountains. This was a remarkable challenge and it took a remarkable man to design and build it: Carl von Ghega, an engineer with mountain-road-building experience who had engineered the Emperor Ferdinand Northern Railway from Brno to Breclav (both now in the Czech Republic). In 1842, Ghega was put in charge of the entire Austrian railroad-building program. Believing that an essential part of this program was a link between Austria and the Adriatic, he traveled to the United States to learn about railroad construction methods, and how they might be applied to finding a way over the mountains. He returned convinced that a railroad over the Semmering was feasible.

The Semmering was a mountain pass that had been used by travelers on foot and on horseback since the Middle Ages. Although it was the lowest of the Alpine crossings, and so remained open longer than others in winter, the pass still rose to more than 3,000ft

CARL RITTER VON GHEGA
Born Carlo Ghega to Albanian parents in Venice, the engineer was granted the title of Ritter ("knight") in 1851 for the astonishing feat of building the Semmering railroad, which many had thought impossible.

THE FIRST RAILROAD ACROSS THE ALPS, 1857

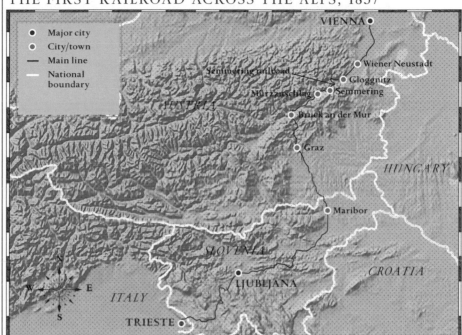

(900m) above sea level, and building a railroad across the range required extraordinary ingenuity and innovation. The Austrian Government backed the project because the revolutions that swept across Europe in 1848 convinced the new Emperor Franz Joseph of its necessity—not only to unite the ends of his disparate Empire as they sought to fragment and affirm their national identities, but also to create employment at a time of economic depression. Establishing a connection from Vienna to the sea became both politically and economically essential.

Ghega chose a route through the Alps that started at Gloggnitz in Lower Austria and ran to Mürzzuschlag in Styria. As the crow flies, these two towns are 13 miles (21km) apart, but the railroad covered twice that distance with its curves and switchbacks, the track running across curved viaducts that arched their way over broad valleys before entering long tunnels. In all, the route required 14 tunnels, the longest of which was 4,600ft (1,400m); 16 viaducts, several with two levels; and more than 100 curved stone bridges. Avalanche sheds were built to protect the line from falling rock and snow on its perilous path along the mountainside. Even with all these structures, the ascent still rose by grades of up to

1 in 40 (2.5 percent), very steep for the locomotives of the time; the curves, too, were sharper than those on other lines. To overcome these difficulties, Ghega also worked to develop engines that could handle them. In 1851, he initiated a competition, similar to the Rainhill Trials 22 years earlier (see pp.27–28), to find the best locomotive. An engine called *Bavaria* won, but when the line opened, it proved unable to haul heavy loads up the steep inclines. A new engine, designed by Wilhelm Freiherr von Engerth, a professor of engineering at Graz University, took its place.

All the construction work was carried out by hand, with the help of gunpowder (the only explosive available at the time). The workforce, which was made up of Germans, Czechs, and Italians, as well as Austrians, was enormous—20,000 people at its peak. Perhaps inevitably, accidents occurred during construction. In the worst single incident—a rock fall in October 1850—14 men lost their lives, and in all, around 700 men died, many from diseases such as typhus and cholera. In one astonishing near miss, the building of the railroad almost changed the course of history. The young Otto von Bismarck, who later unified

THE SEMMERING RAILROAD
Traversing 26 miles (41km) of sheer drops and previously impassable moutains, the Semmering railroad represents an incredible feat of engineering.

Germany and became known as its Iron Chancellor, represented his country at a line inspection. He was nearly killed when a gangway over a ravine broke beneath him—he survived by clinging, cartoon-style, onto a ledge as he fell.

The first freight train traveled over the pass in October 1853 and passenger traffic began the following July. By 1857, the all-important connection between Vienna and Trieste was complete. While the line cost four times its original estimate, it proved its worth as a vital trade link for the Austrian Empire in what turned out to be its declining years. Moreover, the railroad blended so well into the landscape that it has today been designated a UNESCO World Heritage Site, with the double-layered viaducts singled out as a particularly distinctive feature.

NUMBER OF WORKERS THAT BUILT THE

SEMMERING RAILROAD

20,000

NUMBER OF WORKERS NOW BUILDING THE

GOTTHARD BASE TUNNEL

2,000

After the success of the Semmering, other routes were soon being planned through the Alps. A route under the western Alps was proposed in 1848, but the revolutions of that year and the Italian Wars of Unification delayed its progress. Once Italy was unified in 1861, interest revived and a plan was laid to build a line under Mont Cenis, connecting Bardonecchia on the Italian side with Modane in Savoy, which had been annexed to France in 1860 and, more widely, linking Milan and Turin in Italy with Grenoble and Lyon in France.

Going over the Mont Cenis pass was considered out of the question given that it reached a height of 6,827ft (2,081m). However, a temporary rack railroad (see pp.108–109) capable of climbing steep grades, the Mont Cenis Pass Railway, was opened alongside the pass road in 1868. The 50-mile (80-km) line was used to speed up the carriage of mail between Britain and India via the Italian port of Bari. Worked by British locomotive engineers, it was the first ever rack railroad based on the Fell mountain railway system, named for British railroad engineer John Barraclough Fell. The system used a toothed third rail to propel the locomotive up steep hills. However, it proved short-lived and was dismantled in 1871, when the more efficient Fréjus Tunnel replaced it.

The 8½-mile (13.6-km) Fréjus Tunnel, built at a height of 4,000ft (1,200m), was at the time of its construction the longest in the world. Initially, tunneling techniques were still primitive: they involved drilling a hole and placing explosive charges in it. Progress was slow at first. Five years after work had started in 1857, less than 1 mile (2km) had been completed. Then the engineer, Germano Sommeiller, invented a pneumatic rock-boring machine, the pace increased and, with teams working from both ends, the tunnel broke through on December 26, 1870. Thanks to a new method of determining the route known as indirect triangulation, which involved teams of surveyors taking bearings on many points on the mountain, the two ends of the tunnel were less than 2ft (half a meter) out of direct alignment when they met up. The Mont Cenis Railway—the first international railroad linked by a tunnel at the frontier—opened for traffic in October 1871.

The second major tunnel under the Alps was the Gotthard Tunnel, which was also immensely difficult to construct due to the geology of the region. It took from 1871 to 1881 to build and, at 9½ miles (15km), was slightly longer than the Fréjus Tunnel. Work was faster thanks to the use of dynamite, invented by Alfred Nobel in 1867, but it still took an outstanding man, the Swiss engineer Louis Favre, to work out a way of getting through the mountain. Favre devised an innovative way to reach the tunnel mouth, at an elevation of more than 3,600ft (1,100m). He created tunnels that circled around at a gentle grade through the rock, so that the route spirals up the mountain to reach a point above itself. At Wassen, for example, passengers heading south can see a church spire from below then, a few minutes later, they find themselves viewing it from above. These loops add considerably to the journey time—the line between Lucerne and Chiasso on the Swiss-Italian border is 140 miles (225km) long, and around a fifth of the route is made up of loops.

Sadly, Favre did not live to see his creation completed. In 1879, he suffered a heart attack—very likely brought on by the strain of the project—during a tour of inspection of the tunnel, and died at the age of 54. He was not the only one to die: more than 200 tunnel-builders died in accidents—many drowned when drilling hit underground streams, others were killed by rock falls and collisions with the cars used to take out the stone.

After the Gotthard Tunnel opened in 1882, other rail routes were soon carved through the Alps via tunnels including the Simplon, which at 12 miles (19km) became the world's longest when completed

CHAMONIX-MONTENVERS RAILROAD
Following the success of the first mountain railroads, Alpine trains became popular for tourism. The Montenvers rack railroad, which opened in 1908, took tourists up to France's largest glacier, the Mer de Glace.

in 1906, and the Lötschberg, which opened just before the start of World War I. The Swiss developed electric locomotives to pull trains through these tunnels, which would otherwise have filled dangerously with smoke from steam engines (see pp.224–29).

Today, Switzerland is constructing a series of ambitious railroad tunnels through the Alps in order to reduce road traffic over the mountains. The aim is to bypass the old Gotthard Tunnel's slow, winding route, which is already operating at full capacity. To achieve this, the massive AlpTransit project aims to increase rail capacity through these huge new tunnels. The longest—indeed, it will be the longest in the world, surpassing the 14½-mile (23-km) Seikan Tunnel in Japan—will be the Gotthard Base Tunnel. Due to open in 2016, it will have two tunnels, each 35½ miles (57km) long. The new Gotthard rail link, consisting of the Zimmerberg Base Tunnel (for which no completion date has been given) and the Ceneri Base Tunnel (due to open in 2019), along with the Gotthard Base Tunnel, will mean trains can cross the Alps by rail at just 1,805ft (550m) above sea level. This lower elevation will make it possible to create a high-speed link for passenger trains as well as for freight, reducing the travel time between Zurich and Milan from the current 4 hours to 2½ hours. The AlpTransit project also includes the 22-mile (35-km) Lötschberg Base Tunnel between the cantons of Bern and Valais in Switzerland, which opened in June 2007.

All these routes through the Alps represent one of the heroic achievements of the railroad builders. Moreover, they have proved remarkably safe since completion, with accidents a rarity despite the harsh conditions in which the trains operate, especially in winter.

Climbing Mountains

The technology that enables locomotives to ascend steep grades of track was invented in the early days of the railroads. John Blenkinsop's 1811 patent for a steam railroad at the Middleton colliery, Britain, used an engine with a geared cogwheel, or "pinion," that engaged with a line of teeth, or "rack," located between the rails. However, it was not until the 1860s that the system was used on a mountain railroad. Rack-and-pinion tracks maintain traction on grades of up to 48 percent, whereas the steepest incline that conventional ("adhesion") trains can climb is around 10 percent, even with assistance from extra locomotives. These specialized tracks also provide essential braking power to ensure that steep slopes can be safely descended.

The Riggenbach and Locher systems

Various rack-and-pinion systems have been developed and adopted since Blenkinsop's first design.

RIGGENBACH SYSTEM (1863)

The Riggenbach system was the first of the rack-and-pinion systems to be used widely, but its welded "ladder" arrangement proved expensive to maintain.

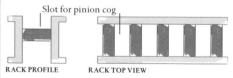

Slot for pinion cog

RACK PROFILE   RACK TOP VIEW

LOCHER SYSTEM (1889)

The Locher system was first used in 1889 and used a horizontal pinion on either side of the rack, enabling trains to climb much steeper grades. It was very stable and allowed cars to withstand crosswinds.

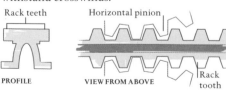

Rack teeth   Horizontal pinion

PROFILE   VIEW FROM ABOVE   Rack tooth

PEAK-SCALING PIONEER
The 3-mile (4.8-km) Mount Washington Cog Railroad in New Hampshire is the oldest rack-and-pinion railroad in the world. It opened in 1868 and climbs over 3,500ft (almost 1,100m) to the summit of Mount Washington, the highest peak in the northeastern US. The line still uses the ladderlike rack created by its founder, Sylvester Marsh, while the engine's tilted boiler is designed to stay level on the steep grades of the track.

How it works

Rack-and-pinion steam engines have one or more pinions, which are powered by the cylinders via connecting rods. Some designs place the pinion centrally on the axle, between the train's wheels, while others mount them on separate axles. Most rack-and-pinion trains have flanged running wheels, and so are capable of running on standard rails. Steam-powered trains pushed their cars uphill, then reversed back down the slope in order to maximize braking power. Today, most rack-and-pinion trains are powered by electric or diesel engines.

THE ABT SYSTEM (1885)

Designed by Swiss engineer Roman Abt and considered an improvement over the Riggenbach system (see below left), the Abt system has a rack with two or three rows of teeth—each row offset from its neighbor—to ensure that the pinion is constantly in contact with the rack.

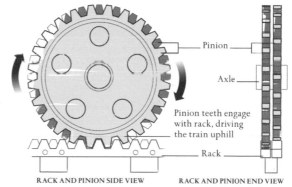

Pinion

Axle

Pinion teeth engage with rack, driving the train uphill

Rack

RACK AND PINION SIDE VIEW   RACK AND PINION END VIEW

The Panama Railroad:
A Deadly Rush for Gold

ALTHOUGH THE PANAMA RAILROAD was less than 50 miles (80km) long, its construction was deadly: as many as 12,000 men may have died building it, from a fatal mix of harsh working conditions and tropical diseases. The railroad created a crucial link between the Atlantic and Pacific coasts of America, however, making it an essential element in the creation of the United States. It was also hugely profitable for its owners and shareholders.

On January 24, 1848, demand for transportation across the Panama isthmus became pressing. On that day, James W. Marshall found gold at Sutter's Mill in California, triggering the first Gold Rush. But reaching California from the east coast of America was a huge challenge. The gold prospectors had three options: to sail 15,000 miles (24,000km) around Cape Horn at the southernmost tip of America, a voyage that took at least 85 days across notoriously storm-ridden seas; to travel 2,000 miles (3,200km) overland across the US on wagon trails, a route that took at least six months and was also fraught with perils; or to sail to the mouth of the Chagres River in what is today Panama and cross the narrow isthmus by dugout canoe up the river and on mules over the hills, to Panama City and the Pacific, an 50-mile (80-km) journey that took up to eight days. By 1848, various canal and rail routes across the isthmus had already been proposed—and abandoned—by La Gran Colombia, the US, and France respectively; indeed, the Spanish had

1849 GOLD RUSH
This guide to the "Wonderful Gold Regions" of California, published in 1849, includes information on the different routes to the goldfields.

THE 1855 PANAMA RAILROAD

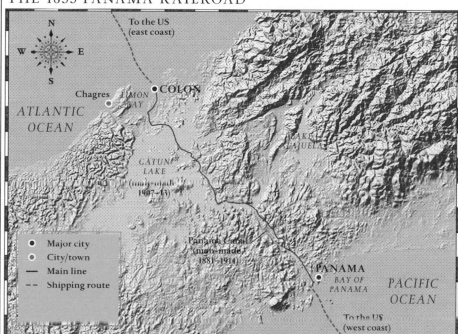

first considered building a canal in the 1520s, before settling for the Camino Real, the overland mule track that was still in use when the first gold-diggers arrived. In 1846, The US Government concluded a new treaty with the Republic of New Granada (Colombia and Panama), which guaranteed the republic's sovereignty in exchange for US transit rights across the isthmus. This paved the way for a transcontinental route. A year later, the US Congress subsidized a mail and passenger steamship service up and down the Atlantic and Pacific coasts from New York to the Chagres River and from Panama City to Oregon, enabling people and goods to reach Panama easily.

New York entrepreneur William H. Aspinwall had won the bid to build and operate the Pacific mail steamships, and, with the onset of the Gold Rush, he set out to build a railroad across the Panamanian isthmus too. To assess the possibilities, Aspinwall traveled to Panama and Colombia with John L. Stephens, a lawyer and writer who had traveled in Central America. They established the Panama Railroad Company, which was granted an exclusive 49-year concession to build a railroad, highway, or canal across the isthmus, as well as 250,000 acres of public land. On the back of this, Aspinwall raised $1 million by selling stock in the company. An astute businessman, he also persuaded the US Congress to pay an annual fee of $250,000 (around £55,000) to transport mail over the isthmus. Meanwhile, the demand for a passenger train had become increasingly evident: by the end of May 1849, 55 ships had landed more than 4,000 passengers at Chagres, all eager to reach California.

The route was first surveyed by US Army colonel George W. Hughes, who was misleadingly optimistic about the railroad's construction. His survey indicated that the terrain would not be hard to traverse: it did not mention the deep swamps, thick jungle, and dangerous hills the route would have to cross. Aspinwall believed the railroad would need to be just 20 miles (32km) long, from the furthest navigable point on the Chagres River to the Pacific Ocean. He contracted experienced American civil engineers George Totten and John Trautwine to build the railroad, but they soon realized the disastrous errors of the survey—for a start, Hughes had overestimated the length of the navigable passage on the Chagres—and withdrew from the contract. However, both were eventually rehired as employees of the company, and the reserved 41-year-old Totten ultimately proved to be the hero of the venture.

Totten and Trautwine were all set to begin work on a route starting from the estuary of the Chagres River, when they found that George Law, the entrepreneur who had won the contract to convey the US mail along the east coast, had bought all the suitable land. They were forced to move the terminus, and found a new site further north at Manzanillo Island. This meant they had to start construction by building a causeway to the mainland, and then build on land known ominously as the Black Swamp. To lay rails over the swamp entailed shipping tons of limestone rock from an abandoned quarry at Bohio, way up the Chagres River, to build a solid base for the tracks. Once on firm land, at the aptly named Mount Hope, they were able to use their first rolling stock—a locomotive and string of wagons. However, less than 1 mile (2km) up the river they encountered further seemingly bottomless swamps and had to sink yet more tons of rocks. Another problem soon emerged: it was pointless building bridges from wood as they decayed within a few months in the tropical climate.

The railroad builders were also unprepared for the weather. Annual rainfall was around 11ft (3.5m), and it rained constantly from June to December. The Chagres River could rise 50ft (15m) in a couple of hours. Not only was this hazardous for men working semi-submerged in the water, it resulted in a climate teeming with tropical

THE VIEW ON ARRIVAL
Passengers eager to cross the Panamanian isthmus arrived
by steamship at the mouth of the Chagres River, shown here,
where they would travel on by dugout canoe, mule—or train.

diseases and insect life. Tarantulas, scorpions, centipedes, wood ticks, and insects such as ants—white, red, and black—could be deadly, and malaria-carrying mosquitoes posed a permanent threat. The swamps were also infested with alligators.

The workers were a veritable foreign legion, turning up from all over the world and often known only by nicknames or numbers on a payroll. Few records were kept and it is not known exactly how many men died. However, it has been reckoned that at one point, one in five of the workers was dying every month. Another estimate suggests that one man died for every railroad tie along the route—around 74,000 men, although this number is implausibly high. Whatever the actual mortality rate, the railroad's doctor, J.A. Totten (George's brother), found it difficult to dispose of the bodies. His solution, wrote historian Joseph L. Schott in his book *Rails across Panama*, was to

> pickle the bodies in large barrels, keep them for a decent interval to be claimed and then sell them in wholesale lots to medical schools all over the world... the bodies brought high prices, and the profits from the sale of the cadavers made the railway hospital self-sustaining during the construction years.

Relations among the international workforce were often fraught. One day, a gang of French laborers stopped work, hoisted the Tricolor, sang the Marseillaise, and refused to discuss their grievances except in their native tongue, which frustrated their Irish foreman. The company chairman, who spoke French, refused to negotiate except in English. He resolved the stalemate by cutting off their rations: the men went back to work, their grievances unknown to this day.

IN ITS FIRST 12 YEARS OF OPERATION, THE PANAMA RAILROAD CARRIED MORE THAN

$750 MILLION IN GOLD

Moreover, the whole region was lawless and subject to banditry at the hands of the Derienni, land pirates who stole the gold transported from California and ruthlessly killed their victims. To combat the Derienni, Askinwall hired Randolph (Ran) Runnels, a famous Ranger who had hung up his guns after a religious conversion, but seen in a prophecy that he would be called upon to take up a mission in a "strange land... with a great river full of demons and monsters." Panama fitted the description, and Runnels set up a mule express business as a front for a vigilante force, called the Isthmus Guard, capable of taking on the bandits. In early 1852, the Guard struck the Derienni while they were relaxing, dancing, and gambling, and hanged 37 of them by the seashore.

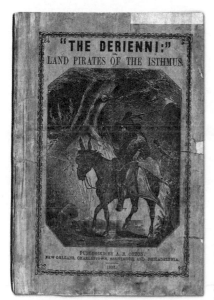

THE DERIENNI
"Graphic Histories" such as this one, published in 1853, fueled tales of the "Robberies, Assassinations, and Horrid Deeds" perpetrated by "Cool Blooded Miscreants" on the trails across Panama.

As if terrible conditions, disease, a restive workforce, and lawlessness were not trouble enough, the company ran out of funds when only 8 miles (13km) of track had been laid; its stock value fell, and construction ground to a halt. Meanwhile, shipping magnate Cornelius Vanderbilt had started work on a rival route through Nicaragua. Fortunately for the Panama Railroad Company, this route was as plagued by problems as their own.

The great turnaround in the railroad's fortunes came in December 1851, when two steamships arrived at the mouth of the Chagres River carrying a thousand passengers desperate to reach California. Hearing the toot of the whistles on Totten's locomotives, they demanded carriage over the existing section—anything to avoid the ghastly mule ride. To deter them, Totten asked an exorbitant fee of 50 cents per mile and $3 for each 100lb (45kg) of baggage. To his surprise, they accepted. Soon, so many people were queuing up to use the railroad that the income from fares enabled Totten and engineer James Baldwin to push rapidly ahead to a limestone quarry, enabling them to transport stone by rail to strengthen the line. The company was able to raise another $4 million in stock on the strength of the demand.

It was not all plain sailing even then, however. In the summer of 1852, many of the workers and their bosses died in a mysterious epidemic. At the same time, renewed banditry forced Runnels to carry out more mass hangings. Then Totten's request to use iron instead of wood for the bridge over the Chagres triggered a dispute with a new company director in New York (Stephens' death had deprived Totten of his major supporter). Totten was fired and the remaining 21 miles (34km) of the railroad entrusted to an engineer named Minor C. Story. Regarded as a boy wonder in the railroad construction business, Story had no idea of the conditions in Panama.

THE PANAMA RAILROAD IN 1859
Local workers propel passengers on a
handcart. Before the Trans-Continental
Railroad was completed, the Panama
isthmus was the fastest route from the
US east coast to the Pacific.

He employed the materials he had used successfully in New
England, but his wooden bridges collapsed in the tropical climate
and he fled, in Schott's words, "bankrupt financially, tarnished in
reputation, and broken in spirit." After a year, Totten was recalled
and recruited more workers, from Europe, India, and China.
Further disasters befell the project, among them a train hitting a
bull on the tracks and toppling into a ravine. Nevertheless, by the
end of 1853 Totten had completed the crucial iron bridge and
the way was almost clear for the run to Panama City.

Around this time, a tragic episode took place among a group of
Chinese laborers, who had proved hard and reliable workers, cleaner
and more sober than the Irish, who mocked them for taking a daily
bath. However, the Chinese relied on a regular supply of opium to

maintain their morale. When an accountant in New York cut off the supply for being too expensive—and criminal—more than a hundred of the desperate Chinese committed suicide, hanging themselves from trees, walking into the water weighed down with stones, or asking Malay laborers to slay them with machetes.

By now the railroad, although still incomplete, was profitable. In 1854, its 31 miles (50km) received over $1 million in fares from more than 30,000 passengers, and on January 27, 1855, the two gangs working from opposite ends of the line joined hands. For 15 years, the Panama Railroad's monopoly of transit across the isthmus – and from the east to west coast of America – made it prodigiously profitable. It paid for its construction within four years on receipts boosted by a scale of charges set up by a group of tipsy clerks: once the railroad was complete, it charged $25 in gold for a first-class fare, making it by far the most expensive railroad journey in the world at the time, mile for mile. In its first 12 years, it transported over $700 million in California gold and more than 500,000 bags of mail without loss, although the cost of maintenance—including the replacement of rotting pine railroad ties with hardwood lignum vitae—dented profits. The Panama Railroad Company's glory days continued until 1869, when the completion of the first transcontinental railroad across the US (see pp.120–27) took business away from it, but 10 years later, the shareholders made a handsome profit when the directors of a French group with plans to dig a canal across the isthmus paid $20 million for the stock.

Totten stayed on as the railroad's chief engineer until 1875, overseeing improvements and maintenance. Immediately after the railroad's completion he had devised a plan for a canal with locks across the isthmus. When Ferdinand de Lesseps, the man behind the Suez Canal, embarked upon his project for such a canal, Totten was appointed chief engineer. Totten also found time to build a daunting mountain railroad in Venezuela. Yet his achievement is recognized only by a modest plaque at the station in Panama City. As Schott writes, "the brief obituary in the *New York Times* stated that he was a retired engineer. It failed to say that he was the man directly responsible for building the first transcontinental railroad the world had ever seen."

TRAIN AT COLON STATION, 1885
A locomotive on the Panama Railroad at the station in Colon, the town that sprung up around the railroad workers' original shanty town on Manzanillo Island.

Crossing America

THEODORE JUDAH WAS A DRIVEN MAN, determined to realize his dream: he believed that the United States needed a railroad running across its vast landmass to connect east and west. The restless, dark-haired son of a clergyman, Judah played the organ as a hobby and came across as rather earnest. Nicknamed "Crazy Judah" in his day, "he was never considered an entirely normal man," according to Oscar Lewis in *The Big Four*. Judah was, however, an experienced railroad engineer who had laid out the route for the spectacular Niagara Gorge Railroad in New York and the Sacramento Valley Railroad, the first in California.

Judah was not the only dreamer: building a transcontinental US railroad had been the ambition of various railroad promoters ever since the first train had chugged down the Baltimore and Ohio Railroad in 1830 (see pp.32–39). Such a railroad was seen not only as economically beneficial but also as a way of bringing together the different regions of the US, which stretched across a land mass of nearly 3,000sq miles (5,000sq km); without the railroad, it is possible that the US might not even have remained united.

It was Judah, however, who almost singlehandedly persuaded Congress to pass a law creating a railroad that would link the existing tracks in the east with California, 1,800 miles (3,000km) away. In 1862, Abraham Lincoln signed the Pacific Railroad Act—an act all the more remarkable given that the Civil War was in full swing and Washington in a state of war. The legislation came with a generous bonus: to ensure the railroad was built, the

THEODORE JUDAH, 1848
Even by his friends, Judah was regarded as a "fanatic"on the subject of the railroads that he dreamed of and built, but he was also resourceful, industrious, and hardworking.

companies that built it would receive government subsidies of between $16,000 and $48,000 for every 1 mile (1.6km) completed, depending on the difficulty of the terrain. Moreover, they would be granted all the land extending 10 miles (16km) to one side of the track.

Once he had seen the legislation passed, Judah traveled to California to survey a possible route through the Sierra Nevada, a rugged mountain range that reaches 14,500ft (4,420m) above sea level. Many doubters argued it would be impossible to traverse the range. Moreover, despite the government funding, Judah still needed more money to finance to project. His breakthrough came at a meeting in the upstairs room of a modest grocery store in Sacramento, where Judah had organized an event to attract potential investors. A group of four enterprising and ultimately very lucky men—Leland Stanford, Charles Crocker, Mark Hopkins, and Collis P. Huntington, all small-town merchants with ambition—decided to back his crazy endeavor. The Big Four, as they became known, founded the Central Pacific Railroad company, which won the contract to build the line heading eastward from California.

Sadly, like many railroad pioneers, Judah did not live to see the fruits of his labors. He fell out with the Big Four over their efforts to extract as much money as they could from the project by fraudulent activity, and headed back to New York. At the time, the only route back to the east coast was via Panama (see pp.110–119), where Judah unfortunately contracted yellow fever and died at the age of 37.

Meanwhile, the Central Pacific encountered difficulties, from winter snow to a lack of funds caused by the corrupt activities of the Big Four. Few local people were willing to work for the company because mining and gold-digging were more lucrative, so thousands of Chinese laborers were shipped across the Pacific to provide manpower. Progress was slow at first, but by 1867, the railway over the Sierra Nevada at the Donner Pass (7,085ft/2,160m) was completed. This was the toughest engineering challenge of the railroad, and

"I am going to California to be the pioneering railroad engineer of the Pacific Coast"

THEODORE JUDAH

subsequent progress on the plains was far easier. When the Civil War ended, in 1865, the Union Pacific Railroad, which had the contract for the other end of the track starting in Council Bluffs, Iowa, began to make good progress. Boosting the Union Pacific's construction, many ex-Civil War soldiers joined the teams of railroad-builders, providing a disciplined workforce supplemented by freed slaves.

Like the Central Pacific, the Union Pacific was deeply corrupt and became a vehicle to enrich its backers, notably Thomas C. Durant, the company's vice-president, and his cronies. Both companies came up with a simple scheme to purloin the public purse: they created separate construction companies, which were given contracts to construct the line at inflated prices, creating vast profits for these companies, which in turn paid out generous dividends to their proprietors—who just happened to be the owners of the railroad companies. In this way, all the main railroad backers became multimillionaires at the government's expense.

Building a railroad in the sparsely populated west of the US required a remarkable level of organization, and as many as 10,000 workers at its peak. The railroad was constructed in stages: first, an advance party surveyed the route; then graders smoothed out the route, shifting huge amounts of rock and earth, laying the embankments and building the bridges; these were followed by the tracklayers, who put down the ties and rails.

The workers lived in camps that moved forward with the railroad. These virtual towns became known as "hells on wheels." They were infamous for their spartan accommodation and saloons—the original of those later portrayed in so many Westerns. Fights were frequent, both with fists and with guns, and the danger of shoot-outs in these townships was constant. The workers also risked attack from Native Americans justifiably incensed at the land grab. The railroad men responded with force to the raids and massacred countless Sioux and Cheyenne people, including women and children, in reprisals, although they established a better relationship with the Pawnee, allowing them free rides on the railroad and establishing an alliance with them against the Sioux.

Groups of workers fought among themselves too. The government contracts had been set up in such a way that the two companies were competing to build the most track, as no meeting point had been specified. At one stage, the two lines passed each other on a mountainside: the Irish laborers of the Union Pacific were blasting rock, which tumbled down on the Chinese workers of the Central Pacific below. Enraged, the Chinese started a fight that was only ended by negotiations between the companies. The Central and Union negotiated a truce over where their respective tracks should meet. This was at Promontory Summit in Utah, where, on May 10, 1869, Stanford and Durant took turns to bang in a golden spike. A momentous occasion, it was marked by celebrations across the US as the news spread rapidly by telegraph. In Chicago, a 7-mile (11-km) parade jammed the streets, while in New York, the event was marked by a 100-gun salute. In Sacramento, 30 locomotives that had been assembled for the occasion tooted their whistles in a tuneless concert. Despite the corruption and construction problems, it had taken only six years for the line to be completed,

THE MEETING AT PROMONTORY SUMMIT, 1869
Samuel S. Montague of Central Pacific Railroad (center left) shakes hands with Grenville M. Dodge of Union Pacific Railroad (center right). The workers' celebratory liquor bottles were removed from some versions of the picture.

GOLDEN SPIKE
The golden spike that marked the "railway wedding" at Promontory Summit, Utah, in 1869—regarded as the day that the eastern and western states of America were united.

rather than the expected ten years—although, in fact, the tracks were not quite continuous across the US until the bridge over the Missouri River between Council Bluffs and Omaha was completed in 1872.

The "railway wedding" at Promontory Summit is still celebrated today in American history as the day on which the country was united. The importance of the achievement cannot be overestimated. A journey that would have taken six months on precarious wagons through a harsh climate—sweating in the summer, freezing in winter—could now be achieved in a few days. It triggered the settlement of the West, and immigrants flowed westward.

Soon, other transcontinental railroads were under construction, both in the US and in Canada. The next two lines opened in 1883— the Southern Pacific (the name of the ocean was used by nearly all the companies), which ran to Los Angeles in California, and the Northern Pacific, which terminated at the other end of the western seaboard in Seattle, Washington. Generous government grants of land stimulated this intense bout of railroad construction, as well as a conviction that the lines would bring profits through the anticipated influx of immigrants.

A fourth line was built without any government support, thanks to the tenacity of a remarkable one-eyed frontiersman, James J. Hill. Known as the "Empire Builder," he was probably the greatest North American railroad-builder. Hill was a strange-looking fellow, lithe and short with a huge nose, a small mouth, and deep-set eyes, one of which he had lost in his youth in an archery accident. He was also a remarkable wheeler-dealer, who for 30 years pursued his vision of building a line that would open up the vast prairies of Montana to settlers and made it possible to export grain to the Far East. Moreover, his line the Great Northern Railway—which in 1893 linked St. Paul, Minnesota, to Seattle, Washington—was built to a higher standard than those of his rivals, with gentler grades and without the tight curves that make trains slow down.

To complete the "set" in the US, the Atchison, Topeka, and Santa Fe Railway crept quietly westward, funded by real estate offices that sold the land that had been granted to it by the government. The railroad, which linked up with existing California lines in 1884, became the most successful of the transcontinental lines by carrying freight through to the port at Los Angeles. Indeed, freight was the mainstay of most of these lines. However, the flow of immigrants proved profitable too, and the lines competed to attract them to the areas served by their routes. The railroads promoted themselves to potential settlers, even opening offices in Europe to attract new migrants so that they could cash in on their vast land holdings. All kinds of dubious claims about the fertility of the land and the mildness of the climate were made to entice desperate people seeking a better life—many of whom gave up after their first winter of heavy snowfall or summer of drought.

Canada, then a British colony, was determined to build a transcontinental line too. This was for commercial reasons, but also to unify the disparate parts of the country—notably British Columbia, which had threatened secession. If anything, the achievements of the builders of the first Canadian line, the Canadian Pacific Railway, were greater than those of their rivals south of the border. As ever, the man in charge, William Cornelius Van Horne, was a remarkable character who took a hands-on approach. He traveled from site to site and was not averse to walking over rickety trestle bridges to show his workers that they must be equally fearless.

It was 2,700 miles (4,300km) from the developed regions of Ontario to the Pacific Ocean, half as far again as the distance covered by the first US transcontinental, and the terrain was certainly no easier. The route of the Canadian Pacific ran to the north of the Great Lakes and, although the initial sections were relatively flat, the granite shelf of the Canadian Shield required considerable blasting through hard rock. There were also two mountain ranges, the Rockies and the

"If he'd of lived, he'd of been a great man. A man like James J. Hill"

F. SCOTT FITZGERALD, *THE GREAT GATSBY*

Selkirks. The line was initially built at a very steep grade of 1 in 22 (4.5 percent) to reach the mountain heights, but these steep stretches were later replaced by spiral tunnels. The tough conditions took a high toll on the 15,000 workers, half of whom were Chinese, and at least 800 men died through want of basic safety measures. It took 13 years to complete the line, which opened without fanfare, on Van Horne's instructions, in 1885.

It was another 30 years before the opening of the next Canadian transcontinental, the Canadian Northern Railway, which took a more northerly route through the mountains. It was built gradually, in sections, with the aim of attracting settlers to the vast prairies of western Canada. The rivalry between the first two lines then prompted the decision to build a third, the Grand Trunk Pacific/ National Transcontinental Railway, built in two sections east and west from Winnipeg. This was the hardest of the three to build, and was in many ways unnecessary, since for hundreds of miles it ran parallel to the first Canadian transcontinental.

As a result of all this frenzied construction, there were no fewer than five transcontinental routes across the US by the end of the 19th century, and the three Canadian transcontinental railroads were all completed by the end of World War I. However, Canada had overreached itself, given its sparse population, and its second two railroads were declared bankrupt soon after their completion—a fate also suffered by several of the American transcontinentals. Nonetheless, these lines continued to operate, providing vital links within their respective countries, and were instrumental in stimulating population growth and economic development: the railroads had conquered the West. Most of the tracks still survive today, playing a vital role in transporting freight.

CANADIAN PACIFIC RAILWAY
A freight train crosses Surprise Creek Bridge in the Glacier National Park, British Columbia. The original bridge, one of several over deep chasms on the line, was wooden. It was replaced by a steel bridge in 1894.

CENTRAL PACIFIC
The crew of a freight train pose for the camera at Mill City, Nevada, on the Central Pacific Railroad in 1883. The train features the classic American "cowcatcher" for clearing the track ahead.

CANADA

Churchill

Dawson Creek

Prince Rupert

Grande Prairie

EDMONTON

Kamloops

Saskatoon

VANCOUVER

CALGARY

Moose Jaw

WINNEPEG

SEATTLE

REGINA

PORTLAND

St. Pau

MINNEAPOLIS

Promontory Point

SALT LAKE CITY

SIERRA NEVADA

DENVER

KANSAS CITY

SAN FRANCISCO

LAS VEGAS

AMARILLO

LOS ANGELES

PHOENIX

SAN DIEGO

FORT WORTH

DALLAS

EL PASO

HOUSTON

PACIFIC OCEAN

Galveston

SAN ANTONIO

Corpu Christ

MEXICO

North American Transcontinentals

The first railroad to cross North America was the Pacific Railroad—a combination of the Union and Central Pacific lines that linked Chicago and California. The fledgling United States of America was unified for the first time, opening up the country to further settlement and exploitation. The success of the first route spawned a profusion of alternatives—as well as three lines across Canada which have together helped the US rail network to become the most extensive in the world.

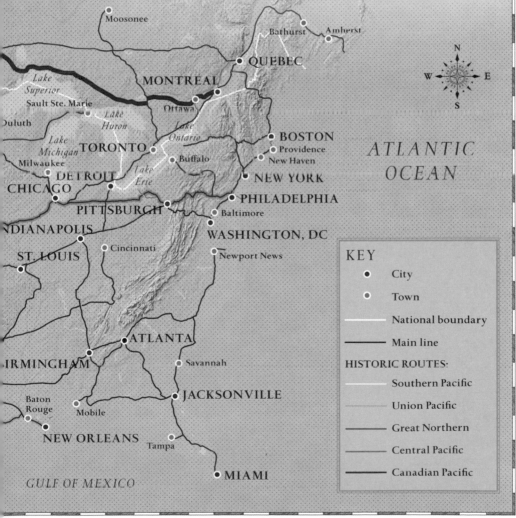

KEY

- ● City
- ○ Town
- —— National boundary
- —— Main line

HISTORIC ROUTES:

- —— Southern Pacific
- —— Union Pacific
- —— Great Northern
- —— Central Pacific
- —— Canadian Pacific

Going Underground

BUILDING A RAILROAD UNDERGROUND was a radical idea, and it took a visionary man to dream up the plan. That man was Charles Pearson, solicitor to the City of London, who realized in the 1840s that something had to be done about the congestion and chaos of the city's streets. The world's first metropolis, London was booming thanks to the Industrial Revolution. In the first half of the 19th century, its population had grown from 1 million to 2.5 million. As it had grown, its traffic problems had become unmanageable, with pedestrians, hackney carriages, and horse-drawn buses all competing for space on the roads.

Pearson realized that central London was crowded with terrible slum housing, but people could not live far from the center as walking was the only way to get to work. The solution, he decided, was to build railroad lines that extended beyond the city boundary; the central section of the railroads, however, would have to be built beneath the city streets, so that houses would not have to be demolished. Pearson hoped these railroads would allow people to move out to far better conditions than existed in the inner city "slums," with gardens and fresh air.

A serial campaigner, who fought for universal suffrage, civic rights for Jews, and penal reform, Pearson is best remembered for his ground-breaking achievement as pioneer of what became known as the Underground. He first set out his idea in 1845, in a pamphlet proposing a glass-roofed railroad down the Fleet valley between King's Cross and Farringdon. Various inventors came up with similar ideas at the time. One plan was to drain the Regent's Canal, which encircled central London, and convert it into a railroad. Another proposal was for an overhead railroad on arches, like those later constructed in New York and Chicago. A third, the most stylish, was for a "crystal railway" around London: a line either above or below street level, combined with a boulevard or walkway lined with shops and houses, and all covered by a glass arcade.

The 1840s were a period of intense railroad construction. A central station for the whole capital was proposed at Farringdon, on the edge of the City of London, but in 1846, a commission ruled this out because of the destruction it would cause in the City. Instead, the mainline stations were located on the edge of the central area. Ironically, this aggravated congestion as passengers pouring off the trains tried to get around the center of the city. Pearson's vision of an underground railroad to link the stations was the obvious solution.

NUMBER OF NAVVIES WHO BUILT THE FIRST UNDERGROUND LINE BY HAND OVER TWO YEARS

2,000

He pushed through his plan, obtaining funds from the City of London as well as investment from existing railroad companies, and the plan received parliamentary approval in 1853. Unusually among railroad pioneers, Pearson did not seek any financial gain from the project.

The Metropolitan Railway Company was founded in 1854, but it was not until early in 1860 that work started on a line to link three of the mainline stations—Paddington, Euston, and King's Cross—with the City, using a new method known as "cut and cover," which involved digging a cutting, laying the railroad, and covering it with a tunnel. This technique was disruptive at ground level and meant that the railroad usually had to follow the line of existing streets. Interestingly, one house that had to be demolished on the route of the line, at 23 Leinster Gardens, was recreated as a façade in order not to ruin an elegant terrace, the open track behind the façade serving as a vent for the steam trains. It became a running joke among postboys to send novices to that address to deliver telegrams.

Although the building of such a line was unprecedented, there was only one major mishap, when the Fleet River burst its banks and flooded the works to a depth of 10ft (3m) in June 1862. Despite this, the 5-mile (7.5-km) line opened in January 1863, only a few months behind schedule, at a cost of £1 million (around £50 million/$81 million in today's money). Sadly, Pearson had died the previous September and missed the sumptuous banquet held at Farringdon station to celebrate its opening.

There had been doubts as to whether people would venture onto this new type of railroad. Not only were the stations gas-lit and dark, but the trains were also hauled by steam engines that belched out smoke and steam, despite being equipped with special condensing

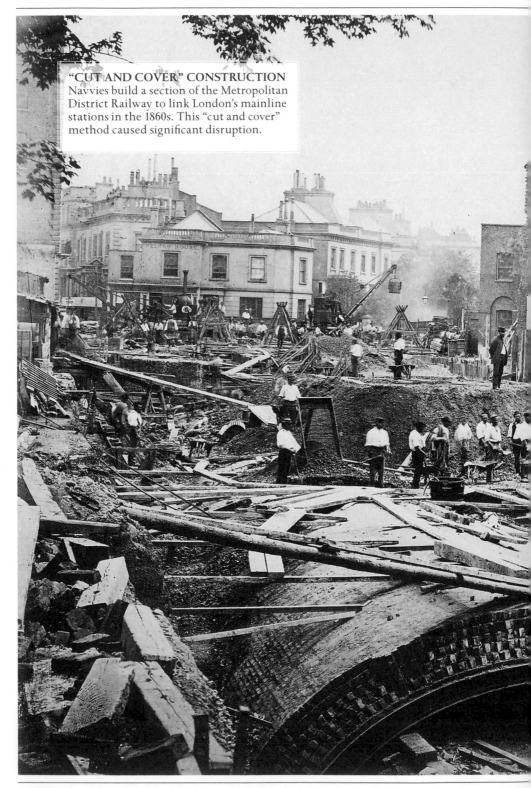

"CUT AND COVER" CONSTRUCTION
Navvies build a section of the Metropolitan
District Railway to link London's mainline
stations in the 1860s. This "cut and cover"
method caused significant disruption.

equipment. The doubters, however, were proved wrong. Londoners had few concerns about trying out this new invention—the world's first subterranean railroad—and 30,000 braved what *The Times* had warned would be "dark, noisome tunnels" to travel on it on the first day, January 10, 1863. The Metropolitan Railway, which gave its name to underground systems all around the world, was an instant success. It provided cheap laborers' trains early in the morning, which proved hugely popular. After the workmen's trains came thousands of office clerks who worked in the City of London and could afford the more expensive fares charged later in the morning. Throughout the day, the trains attracted all types of traveler in three different classes.

The Metropolitan Railway sought to ban smoking at first, on the grounds that the locomotives already created a stuffy atmosphere, but after a complaint in Parliament, the railroad was forced to allow smokers on the trains. Indeed, the air was sometimes so filled with smoke that chemists near the stations enjoyed a roaring trade in Metropolitan Mixture (strong smelling salts), sold as a panacea for passengers overcome by the fumes. In general, though, one of the

UNDERGROUND INSPECTION
Chancellor of the Exchequer William Gladstone (front right) takes a tour of inspection of the Metropolitan Railway, with the company's directors and engineers, in 1862, a year before the line's opening.

"A journey from King's Cross to Baker Street is a form of mild torture"

TIMES EDITORIAL, 1884

reasons people were attracted to the system was that it was safe. No severe accidents occurred during its crucial early days, and indeed very few have happened throughout its history. Within a dozen years of the opening of the first line, no fewer than 70 million people were traveling annually on what had soon become known as the Underground.

Flushed with success, the Metropolitan Railway began to expand almost immediately. It built two extra tracks between King's Cross and Farringdon, called the City Widened Lines, which were then extended farther into the City. Other promoters were keen to get involved, and one was chosen to share the expansion of the network. This was the Metropolitan District Railway (later the District line), run by James Staats Forbes, a rival of the Metropolitan's chairman, Edward Watkin. The two were already in competition when they set out their respective plans to expand the Underground: Forbes ran the London, Chatham, and Dover Railway, while Watkin was a director of its competitor in Kent, the South Eastern Railway.

For the next 30 years, the rival railroads expanded the Underground system rapidly into north and west London. They had rather different conceptions, which is why the Metropolitan line extends far beyond central London into the Metroland developed in the 1920s and 1930s, while the District line only extends as far as the more central Wimbledon, Richmond, and Ealing. Nevertheless, thanks to the dynamism of this duo, the Underground network soon spread well beyond the existing city boundaries, and, wherever the lines were built, new housing soon sprang up.

Despite their achievements, the Metropolitan and District railroads never reconciled their differences and remained in dispute over who should complete the Circle Line, which links nearly all of London's mainline terminals, until its completion in 1884. The two companies ended up running the line jointly, but remained in competition: the Metropolitan ran the clockwise trains and the District the counterclockwise ones. Hapless visitors to London had to choose between the two companies' ticket offices and could end up going the long way round the Circle if they unwittingly bought the wrong ticket.

In 1890, the first deep "tube" line, bored out of the London clay rather than being built by the cut and cover method, was completed. The City and South London Railway ran under the River Thames between Stockwell and the City. As there was inadequate ventilation in the tunnels deep underground, the trains were powered by electricity rather than steam. The line was dubbed the "tuppenny tube" as two pence was the fare for all journeys, and it too was an instant success— despite the fact that the trains operating the service had tiny windows and were universally known as sardine boxes. Other Tube railroads soon followed. Both the Waterloo and City Line and the Central Line, the most successful of the early lines, had opened by 1900.

A series of pioneering promoters continued to develop the Underground. The next was Charles Yerkes, an American with a prison record who had once run Chicago's tramways, and turned out to be a transportation visionary. He brought together several lines, both existing and proposed, under the auspices of the Underground Electric Railways Ltd (UERL); electrified the Metropolitan, District, and Circle lines; and, remarkably, built three deep Tube lines in the space of five years—the Piccadilly and Bakerloo lines, and the Hampstead section of the Northern line. By 1907, all the Tube lines through central London, except the Victoria Line (opened in 1968 and completed in 1972) and Jubilee Line (opened in 1979 and extended in 1999), had been built.

After Yerkes, progress was more in marketing than engineering. Frank Pick, who started working for the network in 1906 and ran it until 1940 alongside chairman Albert Stanley (later Lord Ashfield), created the public image of the Underground. Pick and Stanley were a successful partnership: Pick was an assiduous obsessive, while Stanley was more strategic. Their creation lives on today, a public organization whose achievements have been recognized around the world.

Pick instigated the design of both the famous red, blue, and white roundel that became the instantly recognizable Underground sign, and the innovative and much-imitated map devised by Harry Beck. An electrician employed by the company, Beck created the map, based on electrical diagrams, in his spare time. Pick also commissioned the Johnston Sans typeface for the organization, advertising posters by

UNDERGROUND SIGN
Commissioned by Frank Pick, the Underground logo first appeared as a red disk in 1908. It was later adapted into a bullseye by designer Edward Johnston, and has evolved into the distinctive roundel still in use today.

artists both known and unknown, and cutting-edge architectural designs for new stations. For his part, Stanley persuaded the government to pay for suburban extensions of the system. Together, the pair built up the UERL into a vast organization encompassing almost all the Underground lines and London's buses, and ran the nationalized London Transport from its creation in 1933 until World War II, a period widely regarded as its heyday.

Although the system was neglected after the war and usage fell dramatically, it has revived since the 1970s and flourishes today. Built by people with vision, the London Underground is more than just a transportation system; it is the mechanism that made it possible for London to grow in the way it did. The Underground allowed people to live farther out, as Pearson wanted, but also enabled the center of the city to remain compact and accessible. Today, the underground attracts record numbers of passengers. The network will soon benefit from the addition of Crossrail, running under London in an east–west direction and linking up with all the Underground lines, which will greatly relieve pressure on the system.

Uniquely, the London Underground has become the emblem of the city it serves. Thanks to Pearson and subsequent developers, London became the first city with a railroad beneath its streets, and Britain was once again a pioneer. It was not until the end of the 19th century that other cities started building railroads underground. Budapest, in Hungary, was the second city in the world to develop a metro system, which opened in 1896, more than 30 years after London's. Major cities such as Paris, New York, and Berlin followed in the early years of the 20th century. Now, most major cities boast an underground railroad system and there are nearly 200 metro and subway networks around the world, from Yerevan in Armenia to Los Teques in Venezuela.

Death on the Rails

NOWADAYS, TRAVELING BY RAIL is extremely safe, but this was not always the case. In fact, it took several generations to finally tame the "iron horse." Surrounding this strange and inherently dangerous beast, there were many other hazardous elements. In the words of an old-time railroad man: "Accidents don't happen by accident." Besides the trains themselves—especially their brakes—faulty signaling, defective tracks, human error, and organizational blunders were all culprits. Today, because of the knowledge that train travel is usually perfectly safe, any rail accident attracts an inordinate amount of attention from the media and public. However, no such attention was ever paid to the fate of the earliest railroad workers, who put their lives on the line, literally. In Britain in the 1860s, 800 workers were killed every year, ten times the number of passengers who perished. The number of fatalities among both workers and passengers has declined steadily since then, but working on the railroads, particularly on the track, remains a dangerous occupation.

The sheer metallic brutality of railroad accidents often took more than a physical toll on those involved; they also caused mental anguish, which would now be identified as post-traumatic stress disorder. An early sufferer, British novelist Charles Dickens, described his feelings after he was involved in an accident at Staplehurst in Kent in 1865. At first, he behaved impeccably, nursing the sick and dying. He was, he said, "not in the least flustered at the time," but when he clambered back into his passenger car he could not stop shaking. Recounting the incident later also provoked the same reaction: "In writing these words I feel the shake and am obliged to stop." Dickens never fully recovered from the trauma, remaining frightened of rail travel for the rest of his life, and dying—in a strange coincidence—on the fifth anniversary of the accident.

The only positive outcome of an accident is when it triggers improvement in safety measures or procedures, a process often ghoulishly dubbed "tombstone technology." This was evident as far back as 1842, after the first major disaster in France at Meudon on the line from Paris to Versailles (see p.43). Most of the deaths were caused not by the crash itself, but because the compartments were locked—to deter interlopers who had not paid for their tickets—so passengers were unable to escape from the wrecked train. Thereafter, compartments were left unlocked. Similarly, an appalling crash at Armagh in Northern Ireland in 1889 spurred on legislation that greatly increased the railroad regulator's enforcement powers, and introduced better braking systems. Many of the fatalities in the Armagh crash were children, which heightened the tragedy and prompted such a strong legislative reaction. However, a famous contemporary wit, the Reverend Sydney Smith, suggested that it would take the death of a member of the nobility to really effect change: "We have up to this point been very careless of our railway regulations. The first person of rank who is killed [he suggested a bishop, a breed for whom he had a particular loathing] will put everything in order and produce a code of the most careful rules."

Even today, a safety inspector believes that "the railway gets safer and each incident enables us to improve still further." In the past few decades, installing powered doors, which passengers cannot open, has helped eliminate the problem of people jumping on or off moving trains—a form of idiocy that caused half the deaths and injuries at stations

ARMAGH RAIL DISASTER
On June 12, 1889, a packed train with inadequate brakes slid backward down a steep hill and collided with an oncoming train. It was Ireland's worst railroad disaster—89 people died, mostly children, and hundreds were injured.

reported by British Rail in the 1980s. But tombstone technology has not always been applied. Wooden coaches had long been known to increase the risk of fire in the event of an accident, but as late as 1928, a crash near Bristol, England, involving these old-fashioned coaches resulted in 15 deaths when inflammable gas ignited. Even then, wood-framed coaches were not entirely abandoned in Britain until after 1945. Technological faults can also combine with other conditions to cause tragedy. In India, the deadly Bihar rail disaster of 1981 is thought to have been caused by flash flooding combined with a brake failure. The train derailed and plunged into a nearby river, taking all 800 of its passengers with it. This was one of the worst accidents in railroad history, claiming the lives of an estimated 500 or more people.

Sometimes a train itself may be faultless, but other structural defects can spell doom. Using steel instead of iron has made rails stronger, but the "joint bar" joining rails can be weak. It was a faulty joint bar that caused an accident at Brétigny-sur-Orge near Paris, France, in the summer of 2013, killing seven people. Often, the rail tracks are only as safe as the structures that support them. Most famously, the first rail bridge over the River Tay in Scotland, built in 1878, was not designed to withstand really high winds. It collapsed a year after it was built, plunging a passenger train into the river below. As that memorably bad poet William McGonagall put it:

Beautiful bridge of the Silv'ry Tay!
Alas I am very sorry to say
That ninety lives have been taken away,
On the last Sabbath day of 1879
Which will be remembered for a very long time

2004 TSUNAMI RAIL DISASTER
An earthquake in the Indian Ocean on December 26, 2004 triggered massive tidal waves along the coast. A train in Sinigame on the southwestern coast of Sri Lanka was destroyed.

ON AVERAGE, ONE PERSON IS KILLED PER

3 BILLION

MILES (5 BILLION KM) OF RAILROAD TRAVELED

Some accidents are beyond human control. The deadly earthquake and tsunami of Boxing Day 2004 devastated the coast of South Asia. It also caused the world's worst rail disaster when 1,700 passengers were killed on a coastal railroad in Sri Lanka. Other types of railroad "accident" are all too human, and deliberate. Acts of railroad sabotage have been common, not least in the repertoire of British Army Colonel T.E. Lawrence, also known as Lawrence of Arabia (see pp.288–93). Many saboteurs, such as the French railroad men who wrecked their own tracks in the latter stages of the German Occupation in World War II, are regarded as heroes. Others, such as members of the French Communist party responsible for a crash that killed 21 people in 1947 on the main line between Paris and Lille, are generally considered terrorists. More recently, attacks at stations and on trains as far apart as Madrid, London, and Mumbai demonstrate the vulnerability of railroads to attack by those with evil intent.

Just as pilots are blamed for aircraft crashes, so engineers are always in the spotlight. Fatigue, often after 12 or more hours at work, was the biggest problem before legal limits were introduced. This automatic fatigue syndrome affected not just engineers but other workers as well, such as the overworked signal repairmen who caused the Clapham disaster in London in 1988. Back in 1879, a time when the companies were fighting any attempt to limit working hours, a parliamentary inquiry reported the case of a brakeman who had been on duty for 19 hours and consequently failed to apply the brake on his train. And of course engineers are not immune to personal problems. Many people believe that the worst crash ever recorded on the London Underground, at Moorgate in 1975, which killed 43 people, was caused by a suicidal engineer.

THE LEGEND OF CASEY JONES
In 1950, the US Post Office issued a series of stamps honoring the "Railroad Engineers of America." Casey Jones was one of the engineers immortalized in a stamp, and described as "a railroader's hero."

The most frequent cause of railroad accidents, killing thousands over the decades, is speeding. Nowadays, there are many more controls preventing engineers from speeding, and warning systems to notify them if they do. However, even on modern trains, speeding remains a problem, as seen in the appalling accident that killed 79 people in the summer of 2013 just outside Santiago de Compostela in northwest Spain. One engineer who died while traveling too fast went on to become a cult figure in American folklore and folksong—Casey Jones. A crack engineer on the Chicago Fast Mail, Jones "took his farewell trip to the promised land" when he crashed into a stalled train in the fog in April 1900 while trying to make up lost time. Jones was killed on impact, but he saved the lives of all his passengers and crew by slowing the train at the last moment. As Wallace Saunders put it in *The Ballad of Casey Jones*:

> Casey smiled, said, 'I'm feelin' fine,
> Gonna ride that train to the end of the line.
> There's ridges and bridges, and hills to climb,
> Got a head of steam and ahead of time.'

Thankfully, train crashes are increasingly rare, especially in industrialized countries. In fact, it is car drivers, rather than train engineers, who remain the most likely cause of accidents on the railroads, particularly at grade crossings. In the US, this problem is much worse than elsewhere because there are about 200,000 grade crossings across the country. They account for some 4,000 accidents and 500 deaths every year, nearly all of them car occupants, rather than rail passengers. In three-quarters of the cases, an official report attributed the disaster to "the impatience of the driver of the vehicle involved," a tendency that improvements in the design of grade crossings and modernization of the warning signs can do little to improve. Train engineers have to sound their horn at every road crossing, however minor, which is why in the US trains are heard so frequently. Cumbersome trucks also pose a threat. On a line in rural

France in 1997, a diesel train sliced a slow-moving tanker truck in two, killing 31 people. It was the worst rail accident in France, a country proud of the safety of its trains.

Railroad management has always been a legitimate target for blame, as one 19th-century American lawyer in a case involving a head-on collision remarked: "This is one helluva way to run a railroad." Operations were initially haphazard, because no one knew anything about railroads. As writer-engineer L.T.C. Rolt put it: "When we consider operational methods in the early days of railways the remarkable thing is that there were not more serious accidents." But organizational danger has, if anything, grown over the past few decades, mainly due to the outsourcing of railwork and maintenance, and the increasing prioritization of performance, punctuality, and, of course, cost. These kinds of institutional inadequacies led to the most serious accident of modern times in Europe. In June 1998, 101 passengers were killed after the locomotive on a German high-speed train separated from the coaches, which then derailed. At the time, the Germans were desperately trying to catch up with the French, who had taken the lead in Europe in high-speed rail travel, and had taken short cuts to achieve the right balance of suspension, wheel design, and track flexibility. Deutsche Bahn had been warned of the problem but chose to ignore it, perhaps due to a combination of pride and an unwillingness to delay the introduction of high-speed travel.

Arguably the worst systemic problems in modern railroads—including several fatal accidents—were caused by the privatization of British Rail in the 1990s. It was replaced by 94 separate organizations, many run by inexperienced executives. Track maintenance was also outsourced to a variety of companies, and it was shortcomings in this area that led to a crash at Hatfield, north of London, in 2000. Four passengers were killed and 70 were injured. Investigations into the crash revealed that the cause was a fractured rail, which exposed the failings of the private maintenance companies. Train speeds throughout Britain were reduced to a mere 20mph (32kph) for more than a year, which had repercussions for passengers and train companies alike. As a result of the accident, track maintenance was partially re-nationalized in 2002.

In the 21st century, as in the 19th century, railroads and passengers are vulnerable to both natural and unnatural disasters. However, thanks to modern technology, trains are statistically the safest and most eco-friendly way to travel.

Stopping the Train

In rail's infancy, a train's brakes were simple wooden blocks, called shoes, that were applied to the wheels by turning hand controls at several points along the train's length. As speeds increased, however, a more effective way of braking was required to stop trains over a shorter distance, and various attempts were made to create a brake with one point of control, operated by the engineer. In 1875, a competition was held in Britain to find the best solution. The clear winner of these Newark Trials was the Westinghouse automatic air brake, which was widely adopted in the US. Britain initially used the less successful vacuum brake, but air brakes have since come into standard usage worldwide.

Brake shoes

The first simple brake shoes were soon developed to become "continuous" brakes. In this system, brakes were located on every car and were controlled from the locomotive engine by ropes, chains, or pipes running the length of the train. The wooden shoe was suspended by a lever, or levers, between the brake cylinder and the wheel. As technology progressed, the block was more often made from cast iron, which is still used widely, although modern railroads also use a wide variety of composite materials.

BRAKE SHOE ON NORFOLK AND WESTERN NUMBER 521, 1958

A DANGEROUS JOB
Before brakes were controlled solely from the engine, many railroad cars had brake vans, with hand-operated screw brakes manned by brakemen. The train engineer coordinated the braking with signals from the engine whistle.

Air braking systems

During the 1870s "battle of the brakes," air brakes could bring a train traveling at 50mph (80kph) to a halt in half the time taken by vacuum brakes: the braking distance for Westinghouse's automatic air brake was 777ft (237m), compared to 1,477ft (450m) for a vacuum brake—a considerable safety advantage. Today, air or pneumatic brakes are the standard system used by railroads around the world. These brakes use compressed air to apply a shoe block to the wheel.

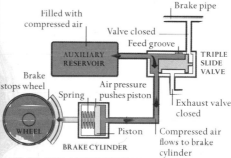

AIR BRAKE APPLICATION

A pump compresses air for use in the system. The engineer controls the air with a triple valve. When this is applied, compressed air is released into the brake pipe and air pressure forces the piston to move against a spring in the brake cylinder, causing the brake shoe to be applied to the wheels.

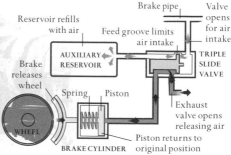

AIR BRAKE RELEASE

When the engineer releases the brake valve, air leaves the brake pipes. As air escapes from the exhaust, a spring in the brake cylinder pushes the piston back, causing the brake shoes to disengage from the wheels. The auxiliary air reservoir meanwhile refills.

The Railroad Experience

TRAVELING ON TRAINS was not comfortable in the early days. Across Europe, the cars were primitive with no amenities, and were based on horse-drawn stagecoach designs. Compartments seated six people facing each other, with doors on both sides. Cars were made up of what was effectively a room supported on two axles, with a basic suspension system. Originally, most of the superstructure was made of wood, which was a fire risk and provided scant protection in the event of an accident.

On the Liverpool and Manchester, and on other early lines, there were different ticket prices. Passengers willing, or able, to pay a little more could obtain a place in the mail car, which had only a couple of comfortable corner seats and offered more privacy. At the other extreme, there was the discount option of riding in a railroad car with sides open to the elements. As passenger travel became more popular, a hierarchy developed. Cheap tickets were available for boxlike cars, which had holes in the floor for drainage, but often no seats. They were used on many railroads throughout Europe and gave people an awful experience of rail travel. Fortunately, these open cars were considered too dangerous, as well as horribly uncomfortable, and were soon phased out. To replace them, compartment-style cars were built for second- and third-class passengers, with less space for people's legs and much harder seats than first-class cars.

Some of the Trans-Siberian Railway services (see pp.180–89) were billed as "luxury", with staff instructed to empty spittoons and keep car temperatures at a balmy 57°F (14°C). In practice, however, customer service was not a strong point. Delays were standard, and at stations peasants scrambled out to cook soup on the platforms, further delaying the train. For everyone else, the dining cars served meals by St. Petersburg time, regardless of the line's seven time zones: toward the east, passengers had to eat breakfast at 2pm, while dinner was served at 3am sharp.

The truly affluent had the option of bringing their own conveyances. The aristocracy simply arrived at the station in their personal horse-drawn coach, which was then lifted on to a flat car and held down with chains, rather as motorists put their cars on some trains today. The upper classes could thus avoid soiling their

THIRD-CLASS SATIRE
A cartoon by the French illustrator Honoré Daumier
showed porters lifting third-class passengers, stiff as
frozen cod, out of a car.

petticoats or pantaloons by sitting on cushions used by the masses,
though even they could not escape the smoke and ashes that
enveloped all early travelers. The dilemma facing passengers when
the weather was hot was whether to open the window and risk
wrecking their clothes from the cinders, or keep it closed and swelter.
An infuriated English social reformer and abolitionist, Harriet
Martineau, reported that sparks had burned no fewer than 13 holes
in her gown during a journey in the United States.

In the early days, the ride was bumpy whatever vehicle class
passengers were traveling in. The springs were weak, or nonexistent,
and the primitive track, made up of short rails, was uneven. Worse, the
couplers between the coaches were not rigid, but involved an
arrangement with chains. Every time the train set off or slowed down,
passengers were thrown about. There was no brake mechanism to
prevent the cars from bumping into one another and despite the
padded seats in the superior cars, complaints from the more well-to-
do travelers about their uncomfortable journey were frequent and

sustained. Luckily, the slowness of the trains reduced the severity of these rough rides, but it was not unknown for the chains to break, leaving some poor passengers stranded. This explains why the last coach of the train was soon eqipped with a red light (called a marker)—its absence would warn signalers that the train was not complete.

The compartment system made it easy for passengers to get on and off, but its main disadvantage was that people could not move within the train, so had no access to toilet facilities or refreshments. When journeys grew longer, trains had to stop at intermediate stations for "comfort breaks"—not an expression used in the 19th century—and meals. That was already the case across the Atlantic, where the design of cars was different from the outset. Interestingly, some of the first American cars were based on canal-boat rather than stagecoach designs. These early cars were more practical and usually more comfortable than their European counterparts. They were longer and were open plan rather than fitted with compartments, which reflected the American ethos of equality. (Equality, of course, only amongst white passengers. Black passengers were segregated until the second half of the 20th century.) Some cars even had seats on the roof, open to the elements, but this idea was soon abandoned.

INSIDE CAR INTERIOR
English painter Thomas Musgrave Joy's 1861 painting
is titled *Tickets Please* and depicts conductors about
to enter a tightly packed third-class car.

American passenger cars were similar to long omnibuses: they could accommodate up to 50 people on two-by-two seats with reversible backs, which were reasonably comfortable. Right from the start, there was a little annex with a hole that opened straight out onto the

COMBINED LENGTH OF TRACK IN THE US IN 1840

2,796 MILES
(4,500 KM)

tracks to serve calls of nature. Relief was only partial. The wheel sets were relatively close to each other, which meant the ends of the coaches swung to and fro, inducing dizziness and even vomiting. The problem was not helped by the fact that there were only four wheels on the early vehicles, but happily this soon changed. Four-wheel vehicles gave way to six and then eight, greatly improving stability.

Novelist Charles Dickens, an experienced rail traveler in Britain, visited the US in 1842 and was pretty dismissive of what he found. Traveling on the Boston and Lowell Railroad, he was dismayed by the lack of class differentiation: "There are no first and second class carriages as with us; but there is a gentlemen's car and a ladies' car: the main distinction between which is that in the first, everybody smokes; and in the second nobody does." He noted, too, that there was a "Negro" car, which was "a great blundering clumsy chest." Dickens particularly took against the American habit of spitting—a fellow author described the central corridor through the train as "an elongated spittoon." Poor Dickens also recoiled when his fellow passengers tried to strike up friendly conversations with him about subjects such as politics, blithely oblivious to the famed English reluctance to talk to strangers and fastidiousness about topics of conversation.

Heating in winter was provided by a pot-bellied stove, which was not only a terrible fire risk when there were mishaps but also according to Dickens filled the air with what he called "the ghost of smoke." The stoves were ineffective, making it too hot for those immediately next to them, but giving no warmth to those further away. Lighting, too, was inadequate. Initially, there were lanterns with candles kept alight by the conductors. The lanterns were placed just above each seat, but gave out little light. Much better kerosene lamps replaced them by the 1860s. They hung from the roof and gave adequate lighting for the whole car, but they too were a significant fire hazard.

Thanks to the open-plan arrangements, American trains attracted hawkers walking along the car, offering things to read as well as drinks and snacks. The first hawkers were self-employed young men who had spotted an opportunity to make money, but later they were officially sanctioned. Many belonged to the gigantic Union News Company. They would pass through the trains offering the day's newspapers, magazines, sweets, soda pop bottles, and cigarettes. They announced their arrival in a falsetto voice, compressing their wares into a single word such as "candycigarettescigars" or "newspapersmagazines." Another British writer, Robert Louis Stevenson, traveling a few years later than Dickens, was much impressed by these young men and was amazed that he could buy "soaps, towels, tin washing basins, tin coffee pitchers, coffee, tea, sugar and tinned eatables, mostly hash or beans and bacon". It was, he noted, much more entertaining than a ride on a British train. But the vendors were not universally welcomed as some ran scams. The favorite was to sell cheaply bound novels for twice the normal 25 cents, with the promise that one of them contained a $10 bill.

The American open-plan model created another difference from Europe: conductors went through the train checking and selling tickets and generally policing the passengers. They were a fearsome bunch and some, who were often on the same train every day or week, became well known to their regulars and even beyond. The doyen of them was Henry Ayers, or "Poppy" as he was generally known, described as "a huge, genial teddy bear of a man, weighing nearly three hundred pounds... [who] hovered over his passengers with benevolent menace." Ayers achieved fame because he had a fierce dispute over the use of the emergency cord with his engineer on the Erie Railroad. The engineer refused to acknowledge that the conductor had ultimate control over the train. By winning the argument, Ayers established railway practice that remains universal to this day and went on to serve the Erie for 30 years. His favorite tale was that he convinced an old lady who had left her umbrella at her station that he had organized to have it sent on

BADGE OF OFFICE
Railroad conductors wore official badges on their uniforms. This one, dated 1880, is from the Baltimore and Ohio Railroad.

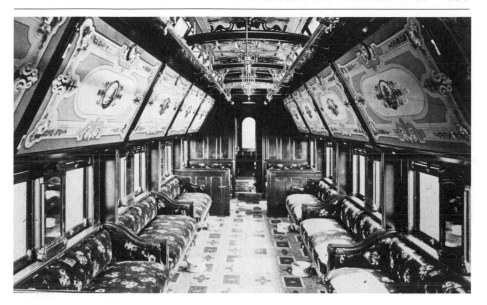

TRAVELING IN STYLE
This luxurious interior of an American railroad
car in 1870 resembles an elegant drawing
room on wheels.

by telegraph. The truth was that lost items were dumped in the
baggage car, so Ayers simply retrieved the umbrella and presented it to
its grateful owner at the next station.

In the early days, train travel in the US was more comfortable than
in Europe because people had the freedom to move about their cars.
There was also an outside area on the last coach, which afforded some
much-needed fresh air in the summer. At first, the connections
between the cars were too difficult for passengers to negotiate safely,
but they soon improved, so people could walk through the whole
train. European cars only began offering the same facility for their
passengers in the last quarter of the 19th century, when trains with
corridors were introduced, but the layout was different. Instead of an
American open plan, the corridor was a passageway at one side of the
car—at first external and used only by rail staff or intrepid passengers,
but later internal. Introducing a corridor marked a significant step
forward in passenger comfort since facilities such as toilets could now
be provided and passengers could have access to a dining car. This
meant that trains no longer had to make intermediate comfort stops.
Although corridor trains became the norm, compartment-type cars
lingered well into the last quarter of the 20th century on some
European local and commuter services.

Turnouts and Sidings

A turnout, or railroad switch, is a track arrangement that allows one set of rails to connect with another. The mechanism consists of a pair of moveable tapered sections of track, known as switch points, which can be pushed into one of two positions, enabling a train to remain on its course or to divert to another line. A common function of a switch is to control access to a siding—a length of track that briefly diverges from a line, allowing a train to be temporarily housed while another train passes by. Sidings can also be used for storing, loading, and unloading cars, and for holding maintenance equipment.

SETTING THE POINTS
Traditionally, switch points were controlled remotely from interlocking towers. However, some still are operated by hand— particularly those that control access to sidings and yards.

Sharing the line

Sidings allow multiple trains to run on the same routes. They are used primarily at stations to let trains vacate the main line so that express services can pass through, or on longer stretches of track to enable freight trains to be passed by faster passenger services. Safety signals ensure that only one train occupies a siding at any one time.

A FREIGHT TRAIN WAITS ON A SIDING, LEAVING THE MAIN LINE UNOBSTRUCTED

Changing tracks

The key component of a turnout is the points mechanism, which was invented by English engineer Charles Fox in 1832. The mechanism is activated by a lever connected to a pull rod, which moves the points from one track to the next. Most points are now electrically operated, but pneumatic versions are also used on some networks, particularly underground lines.

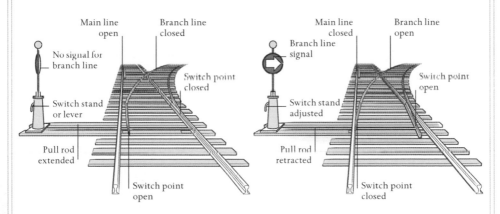

STAYING ON TRACK

The points are set to keep the train on the main line. The lever is thrown, extending the pull rod to slide the moveable rails across the track, drawing the left-hand point away from the track and bringing the right-hand point parallel to it.

SWITCHING TO THE BRANCH LINE

The points are set to divert the train onto the branch line. The lever is adjusted, retracting the pull rod to bring the left-hand point flush with the main track and the right-hand point away from it. The signal indicates the right-hand branch is open.

GREENWOOD SIGNAL BOX, NEW BARNET

ON EARLY SHIFT

THE INTERLOCKING TOWER

An interlocking tower is an operational nerve center from which signals are sent to incoming and outgoing trains and switch points are opened and closed to direct railroad traffic. A single towerman can control an entire lever frame (left) to activate all the relevant signals and switch points.

Temples of Steam

THE LATTER 70 YEARS of the 19th century are commonly known as the railroad age. But it would perhaps be more apt to call the period the railroad station age. Stations were a highly visible presence for travelers, and the first glimpse of a new place for visitors.

Railroads were a revolutionary intrusion into people's lives. They introduced a whole new world of speed, noise, and bustle, all rather frightening to people more used to the gentle clip-clop of horse-drawn travel. But the vast majority of travelers were oblivious to the marvels of engineering that had gone into building the lines, the bridges, viaducts, cuttings, and other structures. The station was the passengers' point of contact with the iron road. Architects had to hit the right note—the stations needed to be solid and reassuring, laying a balm of soothing normality over an unprecedented and therefore alarming experience. Moreover, the buildings had to reflect the importance of the railroad company responsible for the station, and its recognition of the town or society it served.

It took only a decade after the launch of the railroads in 1830 for a quartet of stations to be built worthy of the term "Temples of Steam." The progenitors were the two men who pioneered the modern long-distance railroad, Robert Stephenson and Isambard Kingdom Brunel. Stephenson ensured that both terminals of the railroad he built between London and Birmingham would be worthy of his extraordinary engineering achievement. Stephenson's architect, the eminent Philip Hardwicke, built a noble Grecian-style building at Curzon Street in Birmingham. At Euston in London he constructed a Great Hall of elegant grandeur, with the world's first boardroom upstairs, but unfortunately the heavy Doric arch in front of the station overshadowed the graceful proportions of the Hall.

Brunel could not be outdone by Stephenson. In the early 1840s, his own design for a terminus at Bristol was unexceptional, but 10 years later he worked with a distinguished architect, Matthew Digby Wyatt, to produce Paddington Station in London. Brunel's innovative glass-roofed train shed was complemented by Wyatt's station buildings and 130-room Great Western Hotel, a handsome Renaissance building, and the largest hotel in the country at that time. The hotel and the separation of the station buildings and train shed served as a model for many others – most effectively at St. Pancras in London where the contrasting neogothic hotel and glass-roofed train shed were larger and more dramatic than anywhere else in the world.

Stations were not just the temples of a new railroad age; along with town halls, they showcased the character of a town or city. Railroad companies and their architects continued to emphasize solidity but styles varied wildly as every country reflected its "inner self." Scots went in for Highland style, Germans chose a heavy, Teutonic look, while Spanish and Portuguese stations recalled their ancient Moorish past. Americans went furthest in providing a variety of styles. As railroad historian Lucius Beebe remarked, with only slight exaggeration: "Passengers were set down in storybook settings, Grecian temples, Moorish arches, French chateaux, the tombs of Egyptian dynasts, Turkish mosques, Palladian porticos, Gothic castles and Italian palazzi." The beautiful Antwerp Central station in Belgium is thought to reflect so many different architectural styles as to be unclassifiable; conversely, Flemish architectural style can be seen in the early-20th century Dunedin station in New Zealand. Whatever the style, reactionaries hated the new station buildings. The 19th-century British art critic John Ruskin was appalled at their pretensions to architectural beauty since they represented industry, which, for him, was inherently hideous.

The grandiose Temples of Steam formed only a minute proportion of the hundreds of thousands of stations built around the world during the 19th century. The smaller ones were often delightful examples of local architecture, sometimes standardized by the railway company itself, such as the handsome villas spread over much of western France. In Russia, on the Trans-Siberian Railway, there were five classes of station. The highest class were built of brick and had heated waiting rooms, while the lowest were little more than huts to shelter waiting passengers from the elements.

American stations were generally built in the middle of the city. Thousands of small towns owed their very existence to the railroad, which often ran down the main street. Stations were at the heart of the

"Railway termini are our gates to the glorious and the unknown"

E.M. FORSTER, AUTHOR, 1910

community, full of "retired gentlemen, idlers of all kinds, champion talkers, crackerbarrel philosophers." Although there was powerful opposition to lines and stations near some city centers, in Europe, historic York allowed its ancient city walls to be breached to admit the iron horse, while in Cologne the station abuts the city's historic cathedral. However, if a station was not built in the center it simply resulted in the creation of a new, important part of town. The railroad companies also created their own towns, simply by locating the enormous workshops needed to build and maintain the trains there.

The shape and size of stations outside Europe and the United States often reflected the tastes of imperial masters or European immigrants, notably in Canada where there was a remarkable mix of French and Scottish heritage. In India, the station and its associated buildings such as the engine sheds formed part of an elaborate social and industrial complex, planned by the British colonists. The most impressive example of this is the massive Victoria Terminus (now renamed the Chhatrapati

VICTORIA TERMINUS, MUMBAI
Pictured in 1910, the splendid Victoria
Terminus is modeled on St. Pancras
Station in London. It is still in use today.

Shivaji Terminus) in Mumbai, completed in 1888. In South America, however, architectural roots varied. Argentinian stations, usually built with British money, reflected their financiers' tastes, but in neighboring Uruguay locals built stations in their own style. Stations could also fall prey to triumphalism: when the Prussians retook Alsace and Lorraine from the French in 1871, they imposed their own design on a new station at Metz in Lorraine, complete with statues of Teutonic warriors.

Railroad stations provided the stage for poignant scenes of farewell and reunion, especially during wars. The departure to the Western Front of waves of soldiers from Waterloo Station in London and the Gare de l'Est in Paris left a lasting impression. Then, in 1939, millions of children were evacuated from cities across Europe, helped on to trains by their distraught parents. The emotional power of the station was not lost on film producers either, from the tear-jerking family reunion at a small country station in *The Railway Children*, to the highly charged meetings in a station café of *Brief Encounter*. The world's first film, *A Train Entering a Station*, made by Louis Lumière, depicted a train heading directly toward the camera, causing some of the audience to flee in terror. Artists were inspired by stations, too, such as Claude Monet, who painted a whole series of canvases at the Gare St. Lazare, just below the Parisian café where he and his Impressionist colleagues used to meet.

Stations also created new markets. Two major companies, WH Smith in Britain and Hachette in France, were founded to cater to travelers' needs for reading matter—and so gave birth to the much despised but highly popular "railway novel." However, the glory of a major station could be marred by the inadequacy of the catering. In the half-century before the arrival of special dining or buffet cars on trains, passengers had to rely on refreshments or even whole meals snatched at stations. The owners of these establishments naturally exploited their monopoly. However, Europe led the way in station food, especially France. In Ian Fleming's novel *Goldfinger*, James Bond stayed at railway hotels because, "It was better than an even chance that the Buffet de la Gare would be excellent." Today, the sumptuously decorated Train Bleu restaurant at the Gare de Lyon in Paris is justly famous, as is the Oyster Bar at Grand Central Terminal in New York City.

By the end of the 19th century, station architects—and the companies behind them—were confident enough not only to design afresh, but also to use architecture to express a political, social, or national vision. The modernist station with exciting clean lines designed by Eliel

Saarinen for Helsinki, Finland, in 1919 announced not only the arrival of the modernist movement but also proclaimed Finland's newly declared independence from Russia (1917). And after World War I, the French gloried in local Norman and Breton styles in a number of provincial stations, while at Perpignan in French Catalonia they erected a statue of Spanish surrealist artist Salvador Dalì. Even the new station at Milan, Italy, in 1930 was Mussolini's statement of fascist grandeur.

World War II destroyed many great stations, but postwar reconstruction programs included the last true Temple of Steam, the Roma Termini in Rome, Italy. As the motorcar usurped the train, however, stations became neglected. The train services in many countries were scaled back, with some routes and stations permanently closed. A few stations were rebuilt but others were demolished—the most tragic victim was Pennsylvania Station in New York. The worst officially sanctioned railroad vandalism occurred when the Belgian railroads carved a cutting through the heart of Brussels to link the stations to the north and south, inflicting a permanent wound on the city. However, some disused stations were converted for other worthy purposes, such as the Gare d'Orsay in Paris, which now houses the Musée d'Orsay, a national art gallery, or Manchester Central Station in Britain, which was transformed into a convention and concert venue. In the US, some great union stations, such as St. Louis, survived without any trains by becoming shopping, hotel, and entertainment complexes.

Today, many of the stations that survived destruction during the motorcar boom of the 1960s are now flourishing, with their architectural heritage preserved. St. Pancras Station in London is a splendid example of the renaissance not just of trains but also of their stations. Until the end of the 20th century, it was best known as a decaying architectural masterpiece, which had been saved from demolition in the 1960s, but housed only a derelict hotel and a grimy set of platforms. Today, thanks to a substantial renovation completed in 2007, it is a world-beater. In Spain, the magnificent Atocha station in Madrid underwent a rather unique rebuild; originally designed in collaboration with French engineer Gustave Eiffel, during the 1990s its striking main hall was converted into a huge botanical garden, complete with turtles. Stations such as St. Pancras, Grand Central Terminal in New York, Kyoto Station in Japan, the Berlin Hauptbahnhof in Germany, and Toronto Union Station in Canada are no longer used simply by travelers passing through, but have become destinations in their own right.

GRAND CENTRAL
Seen here in 1930, the majestic concourse at Grand Central Terminal in Manhattan looks remarkably similar today, thanks to a loving restoration.

Railroad Signal Telegraphy

The telegraph transformed railroad signaling, making it possible for train operators to send messages ahead of trains for the first time. American inventor Samuel Morse devised the earliest experimental telegraph, and the Cooke and Wheatstone needle telegraph, a later model, first entered commercial use in 1838 when it was adopted by the Great Western Railway in Britain. The system gained wider acceptance after its dramatic role in apprehending British murderer John Tawell in 1845: he had been seen boarding a train at Slough, and this information was telegraphed ahead to Paddington Station, where he was arrested. In 1844, Morse's telegraph transmitted the words "What hath God wrought" from Baltimore to Washington, DC, and brought about a revolution on the US railroads.

The needle telegraph

Inventor William Fothergill Cooke and scientist Charles Wheatstone patented their five-needle telegraph in 1837. It consisted of a receiver with needles that were moved by electromagnetic coils to point to letters on a board. Each letter was communicated via two currents flowing down two wires, causing the receiving telegraph's needles to swing to the left or right. Six letters were omitted—C, J, Q, U, X, and Z—a limitation that caused confusion when identifying the murderer Tawell (see above), who was described as wearing a "KWAKER" (Quaker) coat.

COOKE AND WHEATSTONE FIVE-NEEDLE TELEGRAPH (1837)

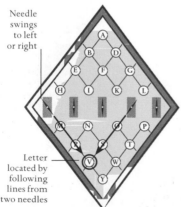

Needle swings to left or right

Letter located by following lines from two needles

READING THE MESSAGE
The five-needle telegraph was popular with users as it did not require any knowledge of codes: two of the five needles pointed to one of 20 letters laid out in a diamond pattern on the receiver to spell out words. Over time, the system was simplified to two needles and then to a single needle, mainly to reduce the cost of replacing wires as they deteriorated. However, systems with fewer needles required complex codes, so telegraphy became a specialized job.

Morse's telegraph

Invented in 1835, Morse's first telegraph used a pencil point attached to an electromagnetic pendulum. His partner Alfred Vail suggested using a lever and armature to print a code of dots and dashes—the precursor to Morse code— sending the message "a patient waiter is no loser" in 1838. This system, patented in 1840, was soon adopted in the US both for railroad signaling and general use, with lines built alongside new railroads. It was cheaper and simpler to use than the needle telegraph, especially once adapted to an audio system.

MORSE KEY SOUNDER (1875)
The Morse telegraph had one wire and an earth connection, and used a single current switched on or off to send dots and dashes.

REPAIRING THE WIRES
A lineman repairs a telegraph line at Billericay, Essex in the UK in 1932. The telegraph remained integral to railroad signaling, and was still in use as late as the 1970s.

Monopolies and Railroad Barons

ACCUSTOMED AS WE ARE TO THE NOTION of national railroads and large rail monopolies, it is hard to imagine today that the early railroads were made up of hundreds of small companies, each serving its own limited local area. Local entrepreneurs created these railroads, largely from a desire to improve transportation for their own goods rather than as a result of any grand vision. However, as the railroads expanded, they became more ambitious in scope, and economies of scale became evident. Not only was it convenient for people to travel long journeys on one train, it was also more profitable for the train companies to operate in this way. It was relatively cheaper to carry goods over a longer distance, as loading and unloading them was expensive.

Lines gradually grew longer. In Britain, the London and Birmingham Railway was the first railroad more than 100 miles (160km) long, and a civil engineering project far larger than any that had gone before. Robert Stephenson, son of George Stephenson (see pp.22–29), was appointed chief engineer in 1833 and the railroad, London's first main line, opened five years later in 1838. Other, longer, railroads soon followed, such as the Great Western from London to Bristol, constructed by engineer Isambard Kingdom Brunel between 1835 and 1841 at a cost of $32.5 million (£6.5 million), and the London and Southampton, completed in 1840. In the United States, both the pioneering railroads—the 136-mile (219-km) Charleston and Hamburg, opened in 1833, and the Baltimore and Ohio, whose first section ran from 1830—were ambitious projects that extended far inland from the Atlantic Coast port cities they served.

The men who managed the railroad companies were generally enterprising and progressive, and could see the benefits of far-reaching railroad networks. As the railroads became established, these men dreamed up ever-grander plans, and railroad promoters raised financing for their construction. These longer railroads began to sprout branches, and the companies grew. Given the advantage of large networks, it was inevitable that railroad companies should come together first to provide joint services. However, even as they extended their reach, the railroad companies remained distinct until

THE BALTIMORE AND OHIO RAILROAD
Among the first American lines, the B&O was also
part of the heyday of the railroad barons and the
battles between the first large companies. Pictured
here is the early B&O locomotive *Atlantic* in the 1830s.

competition between them made the creation of monopolies
attractive and they began to merge—a process that accelerated
when railroads fell on hard times, and rivals took over their
struggling competitors.

Consolidated railroad companies soon became the biggest
enterprises in the world. In many respects, their development prefigured
the growth of capitalism. Before the railroads, there were no large
companies in any industry, with the exception of quasi-government
organizations such as the Dutch East India Company. Factories were
still just small buildings where goods were made, owned by local
entrepreneurs and supplying their immediate area. These firms
employed local people and were part of the local community.

The railroads were a totally new and different sort of enterprise.
To begin with, they stretched dozens and later hundreds of miles,
taking them across many different localities. Then, they demanded a
new style of management with complex organizational structures.
Furthermore, the companies needed bosses with vision and great
ability, able to see the bigger picture as well as being industrious. In
Britain, already off to a head start with the railroads, the first larger-

scale, consolidated, companies began to emerge. The new companies were powerful and self-important, as testified by the grand stations they erected in major cities (see pp.152–57). The first mega-company to spring from the dozens of smaller railroads was the Midland Railway, created by entrepreneur George Hudson (see pp.54–55). In 1844, Hudson created the Midland by consolidating three lines: the North Midland Railway, the Midland Counties Railway, and the Birmingham and Derby Junction Railway, all of which converged at Derby. The new, combined railroad created the core of a line that ran all the way from London to York, offering a more convenient service to its passengers, with fewer changes of train. Hudson had also built the Newcastle and Darlington Junction Railway, which he now linked to the rest of his network via York, so that he had control of more than 1,000 miles (1,600km) of railroad. He continued to consolidate the network throughout the 1840s by taking over other smaller Midlands lines. In 1842, Hudson cleverly created the Railway Clearing House, an organization that enabled all the railroad companies—of which there were now more than a hundred in Britain—to collect revenues from each other when passengers used more than one company's trains on a journey. Until then, passengers had been obliged to change trains and buy a new ticket at each stage of the journey. Sadly, for all his entrepreneurial genius, Hudson turned out to be a fraudster. The ticket-sharing plan, however, long outlasted his downfall and the collapse of his railroad empire.

Despite the obvious advantages of consolidation, British politicians were reluctant to let the railroads merge, fearing that monopolies would exploit the public. However, the financial weakness of some railroads forced the government to accept the idea, especially at times of economic downturn, so throughout the mid-19th century, companies continued to merge.

One of the merger beneficiaries, the London and North Western Railway (LNWR), was, for a while the world's biggest company, employing 15,000 people at its peak. Its brilliant manager, Captain Mark Huish, was a former Indian Army officer. He shrewdly negotiated takeovers, managed the company in an innovative manner, and introduced novel accounting methods essential for the company's size. The LNWR was created in July 1846 by the merger of the Grand Junction Railway, the London and Birmingham Railway, and the Manchester and Birmingham Railway. This created a network of 350 miles (560km). The core route connected London with Birmingham, Crewe, Chester,

Liverpool, and Manchester. Although it covered less ground than rivals such as the Midland, the LNWR offered the best route between London and the main towns of northwest England.

Once it was in a position of strength, the LNWR bullied rivals into forced mergers or disadvantageous deals to run on their tracks. But the bully-boy tactics did not always work. Two small railroads that combined to run on a shorter route between Birmingham and Chester than their big rival were warned off by Huish: "I need not say if you should be unwise enough to encourage such a proceeding, it must result in a general fight..." To his alarm, they resisted the threat, fought a three-year battle through the courts, and, surprisingly, won. However, this was an exception. For the most part, the big bullies effectively quashed their smaller rivals.

In continental Europe, major networks were soon created from the plethora of companies that had started running rail services. In France, six big regional companies were formed with government encouragement between 1858 and 1862. The most ambitious, the Paris–Lyon–Mediterranée (PLM), soon extended over the border to Switzerland and Italy, creating an international network. The Rothschild banking family owned the PLM, along with a second French company, the Nord, and it looked for a time as if the Rothschilds would establish themselves as the dominant force in European railroads. At one point they also controlled the Austrian Südbahn, which included the Semmering Railway (see p.102–107), and had concessions on various Italian lines. After unification, however, Italy prevented the Rothschilds from expanding their railroad empire and effectively blocked the creation of "the biggest international railway company that had ever been seen." Instead, four big Italian companies were formed, but the poverty of the country and its difficult

PARIS–LYON–MEDITERRANÉE
One of the big six railroad companies in France, and one of the first international companies, the Paris–Lyon–Mediterranée promoted its services to tourists with posters designed by artists of the day.

terrain, with the Apennines running down its spine, made it hard to make a profit. The nation suffered a railroad crisis every few years and, unwilling to allow foreign takeovers, Italy in 1905 became one of the first nations to nationalize its railroad system. In the early days of Swiss railroads, meanwhile, the plutocrat Alfred Escher (see p.55) more or less oversaw the explosion in railroad construction from the 1850s, and became so powerful that he was nicknamed "King Alfred."

The American railroads soon eclipsed all others in scale. The distances to be covered were huge, so big companies such as the Erie and New York Central Railroad and the Pennsylvania Railroad emerged early in railroad history. When the first transcontinental was completed in 1869 (see pp.120–25), the Union Pacific and Central Pacific railroads became the railroad giants. After them, a series of American railroads earned that accolade as railroad barons consolidated the network.

By 1900, seven companies controlled most of the US railroads. Many of their proprietors became infamous, among them Jay Gould and his son George, J.P. Morgan, Edward Harriman, and the Vanderbilts. William, the younger Vanderbilt, demonstrated the attitude of the new railroad barons to their passengers when asked by a reporter why a popular fast train was no longer operating: "The public be damned! ... I don't take any stock in this silly nonsense about working for anybody but our own," he is alleged to have replied. Daniel Drew, one of the first of these "robber barons"—a term applied by *Atlantic Monthly* in 1870 to the new breed of capitalists—was also one of the

OUR ROBBER BARONS.

OUR ROBBER BARONS,
PUCK, 1882
In this cartoon from satirical magazine *Puck*, "R. Road Monopolist" Jay Gould, William Vanderbilt, and other barons divvy up their spoils by "Castle Monopoly."

"We hear now on all sides the term 'Robber Barons' applied to some of the great capitalists"

ATLANTIC MONTHLY, 1870

biggest rogues. As company treasurer of the Erie, Drew agreed several times for the company to borrow money against newly issued shares, and then used his position as an insider to profit from the trade in these shares—insider trading was not regulated at the time.

J.P. Morgan was the most successful of these entrepreneurs, forging a railroad empire that stretched across the US by taking over ailing companies and reorganizing them. Unlike Drew, Morgan actually improved the railroads he acquired. So too did Harriman, who became known as the "greatest rail baron in America." His background, as with most of these moguls, was modest: he started out as a messenger boy, but made money on the stock exchange and invested in railroads. He boosted his assets by purchasing rolling stock, improving tracks, and establishing good management. His first major acquisition was the Illinois Central Railroad. After the Panic of 1893—the last of a series of major 19th-century recessions caused by boom and bust in the business cycle—he added the massive Union Pacific to his set. Harriman made it profitable by straightening out bends in the track—many added unnecessarily by the builders to take advantage of government subsidies—and reducing grades, and it became a moneymaker. By the turn of the century, Harriman controlled more track than any individual in the history of American railroads. He was the baron of barons.

All too soon, the advent of the motorcar brought about the decline of the railroads, and empires were broken up or bailed out by the government. In the US, however, one last pair of barons emerged in the 1920s: the Van Sweringens. This strange pair of reclusive brothers were property developers who bought a railroad for its development potential and ended up with an empire that included the Erie, the Chesapeake and Ohio Railroad, and the Pere Marquette Railroad, which operated a series of lines in the Great Lakes area. Following the 1929 crash, however, their empire was broken up even more quickly than they had built it up. With this, the era of the big railroad barons finally came to an end.

Building Bridges

As the railroads expanded, railroad bridges were erected to enable lines to follow the most direct routes possible. Engineers devised a range of architectural and engineering strategies for overcoming the obstacles of local geography—from vaulting brick viaducts spanning valley floors to iron suspension bridges that dangled tracks above rivers and gorges. Concrete and steel are now the materials of choice, but the ingenuity of modern designs is no less impressive than the elegant blends of function and form deployed by the early railroad builders.

The Tangiwai disaster

New Zealand's worst ever rail bridge disaster occurred on Christmas Eve, 1953, when 151 lives were lost in the catastrophic failure of the Whangaehu River bridge near Tangiwai, in the center of the North Island. A mudflow from a collapsed dam destroyed a supporting pier of the beam bridge (see panel, opposite), which gave way under the weight of an express train.

ENGINEERING MARVEL
At the time of its completion in 1890, the Forth Bridge in Scotland was the longest steel bridge in the world. Its three 330ft-(100m-) high trapezoid cantilevers support the railroad at a height of 151ft (46m) for the 1.6-mile (2.5-km) crossing of the Firth of Forth sea inlet.

Types of bridge

Different types of railroad bridges have been developed to meet geographical and economic constraints. While there are multiple variations on—and combinations of—each design, the basic types of bridge can be grouped into four categories: beam, arch, cantilever, and suspension.

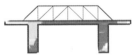

THROUGH-TRUSS BEAM

DECK-TRUSS BEAM

BEAM BRIDGE
The simplest of all bridges, a beam bridge, consists of a "beam" or girder laid across a gap, supported at each end by piers and often strengthened by a truss. A through-truss design uses iron or steel struts joined together in triangular sections to form the load-bearing superstructure, which carries the railroad track beneath. A deck truss uses the same arrangement of struts, but supports the beam from below.

THROUGH ARCH

CANTILEVER

ARCH BRIDGE
A classic arch bridge supports the railroad from below, but in the more sophisticated through-arch design, the top of the arch rises above the deck, suspending the railroad from vertical cables or struts.

CANTILEVER BRIDGE
Built from sections that are supported at one end only, cantilever bridges have the advantage of not needing "falsework" (temporary supports) in construction, so are ideal for wide crossings.

CLASSIC SUSPENSION

CABLE-STAYED SUSPENSION

SUSPENSION BRIDGE
Well suited to long crossings in exposed, windy locations, suspension bridges make use of cables to support the deck. The classic design stretches a cable laterally between one or more towers, with vertical suspension cables to support the deck. A cable-stayed bridge suspends the deck directly from one or more towers via a series of angled suspension cables.

The Pullman Phenomenon

THE NAME OF ENGINEER GEORGE PULLMAN is still used to refer to the sleeping car he developed for rail passengers. He was the American "genius of the bed on wheels," who destroyed all his competitors to establish a monopoly in the US by the end of the 19th century. As with many such legends, the original idea was not his, but Pullman made it his own.

It was inevitable that the first sleeping cars would be developed in the US rather than in Europe, as journey times were longer and trains were slower because of the sharper curves and higher grades of the tracks. At first, trains simply stopped for the night and passengers stayed in local hotels and inns. This was clearly both unsatisfactory and inefficient, so the first sleeping car was introduced on the Cumberland Valley Railroad in Pennsylvania in 1839. Unfortunately, it did not provide a comfortable night's sleep. The sleeping accommodation consisted of a couple of cars, each with four sets of three-tiered berths. They were no more than hard boards without bedding or mattresses, which were then folded away during the day.

A few years later, the New York and Erie Railroad devised an equally uncomfortable solution. Two cars, known as "diamond cars" because of the shape of their windows, were equipped with iron rods that could be used to link facing seats to create a basic bed. The cushions were made of horsehair cloth that penetrated all but the thickest clothing and were invariably infested with all kinds of ravenous insect. As if that were not bad enough, the condition of the track, with its short, badly laid rails, made the experience akin to "sleeping on a runaway train," according to one early passenger.

By the 1850s, matters had begun to improve. Several railroads advertised improved sleeping accommodation based on an idea by Webster

GEORGE MORTIMER PULLMAN
Born in New York in 1831, Pullman achieved success at a relatively young age. He had developed his first railroad sleeping car before the age of 30.

Wagner, a stationmaster on the New York Central. He went on to found the Wagner Palace Car Company, a rival of Pullman. Wagner developed the idea of a coach with a single tier of berths and bedding closets at each end, a definite improvement on all its predecessors. Several other

THE VALUE OF PULLMAN'S PATENTS IN 1875

$100,000

(TODAY, $2 MILLION, OR £1.2 MILLION)

competitors emerged, building sleeping cars to a variety of designs. But it was Pullman who transformed nighttime travel on trains, making it not only comfortable but even respectable.

Like many railroad entrepreneurs, Pullman had already been successful in a different enterprise—that of moving houses (not their contents, the buildings themselves). Several low-lying houses in New York State were in the way of a planned extension to the Erie Canal, so Pullman, along with his father, contrived to move them to higher ground by putting them on wheels. Pullman then moved to Chicago, where he started a fledgling railcar business in 1858, building two cars for the Chicago and Alton Railroad that provided upper and lower sleeping berths.

The radical aspect of Pullman's design was that the upper berth was suspended from the ceiling by ropes and pulleys. When not in use, it could be hauled up to the roof, leaving plenty of seating space during the day—unlike older designs, in which the bunk was fixed and made life uncomfortable for passengers when they were not in bed. Curtains around the berths created privacy, but these cars were still crude affairs. Candles provided lighting and a wood stove generated heat, but both were a great fire risk given the large number of flapping curtains. Each car accommodated 20 people, and blankets and pillows were provided, although not sheets. The experiment was successful. Soon the cars, which ran between Bloomington, Illinois, and Chicago, filled up every night. The only problem was dirty footwear. The conductors had to convince the male passengers—who were the majority of travelers at this time—to take off their boots at night so that they didn't soil or damage the bedding. There was a great reluctance to do so, presumably because of fear of theft, so for many years every Pullman car carried notices politely requesting, "Please take off your boots before retiring."

PULLMAN PALACE CAR
Nicknamed "palace cars," these sleeping cars were modeled on the packet boats that a young George Pullman had seen traveling on the Erie Canal.

In 1861, in association with local authorities, Pullman supervised the raising of many of Chicago's buildings in an attempt to improve the city's street sewerage system. His most spectacular feat was the four-storey Tremont House Hotel, the tallest building in Chicago at the time. The hotel was in line for demolition, but Pullman devised a clever method to save it. He put the whole building on 5,000 jacks, then a team of 1,200 workers turned the screws 180° each time Pullman gave the signal. They successfully raised the hotel by 6ft (1.8m), while the house band continued playing and the hotel guests happily ate their lunch.

Perhaps raising the Tremont House Hotel helped to inspire Pullman to create a moving hotel for rail passengers. In 1863, he set about building what he claimed, with some justification, would be the finest and most luxurious sleeping car ever. Pullman had realized that he needed to attract the rich and famous, and his efforts were rewarded with good fortune. His new car, called the *Pioneer*, cost $20,000 to build, perhaps four times the cost of any other railroad car at the time. Mary Todd Lincoln, the First Lady, saw it on a visit in early 1865 and was enchanted by the style and elegance of the luxurious train, with its hand-carved seats and panels, and thick pile carpet. When she was called upon to organize the funeral of her assassinated husband a few months later, she remembered the car, and it was used as the hearse for the funeral procession along the Chicago and Alton Railroad.

The *Pioneer* attracted national attention, and soon Pullman began building cars for other railroad companies. Before long, many other railroad companies, such as the Michigan Central, the Burlington, and the Great Western, were attaching luxurious Pullman cars to their trains. Pullman's company also sold the tickets for the berths, which cost 50 cents more than the railroad's own sleeping accommodation, but were far superior. Passengers were carried in comfort in Pullman sleepers to almost every part of the US. The cars were all built to the same design, equipped for both day and night travel, and served by Pullman's employees. It was a great business model, and highly profitable as Pullman did not have to pay any of the costs of hauling the trains or using the tracks.

Oddly, despite the luxurious surroundings, the Pullman cars were inferior to those in Europe in one key respect: they did not provide the same level of privacy. Pullman's design was open plan, with makeshift folding seats, pull-down berths, only curtains for privacy, and nothing to stop a loud snorer from keeping the whole car awake. In Europe, compartments were retained, containing up to six beds on three levels, although a few open-plan cars were introduced for third-class passengers, notably on the Trans-Siberian Railway (see pp.180–89). The open-plan design survived in the US up until the second half of the 20th century, and can be seen in several famous scenes in the Billy Wilder classic comedy *Some Like It Hot*, starring Marilyn Monroe. Although the cars were popular, some commentators disapproved of such close-quarters living arrangements.

In 1867, Pullman developed a combination sleeping and eating car, with a kitchen at one end and removable tables set between the seats at mealtimes. Although this was not the first time that meals had been provided on board trains, the quality, Pullman's selling point, was undoubtedly superior to anything that had gone before. Sugar-cured ham was 40 cents, a Welsh rarebit 50 cents, and steak with potatoes just 60 cents—cheap even at the time. Pullman was a great publicist and introduced his first hotel car on a journey

"It is considered tolerable that [women] should lie with the legs of a strange, disrobing man dangling within a foot of their noses"

KARL BAEDECKER, GUIDEBOOK AUTHOR, DESCRIBING PULLMAN'S COACHES

LINCOLN'S FUNERAL TRAIN
Hundreds of thousands of people lined the railroad as Lincoln's funeral train passed by, providing Pullman with excellent publicity and many new customers.

around the eastern US from New York to Chicago that took seven days. Next, he put self-contained dining cars on trains. The first was called *Delmonico*, after an eminent restauranteur of that name. Pullman tried it out on the Chicago and Alton, his home railroad, on which he always tested his ideas. It was another great success, and Pullman subsequently both built dining cars for other railroads and operated them himself on some lines. The selection of meals for sale on the best trains was sumptuous. The hotel cars on the Chicago-Omaha service in the 1870s offered a choice of 15 seafood and fish dishes, together with 37 meat courses, including a huge variety of game. One only wonders how often the waiters had to say: "Sorry, that's not available today."

While Pullman's customers greatly enjoyed the quality of service, it was a terrible idea to die on one of his trains. Pullman ruled that if a passenger passed away, the corpse had to be put off at the next station, regardless of whether the town had an undertaker, leaving the traveling companions of the deceased to deal with the situation. Fortunately, most of the attendants were more compassionate than Pullman, and would ensure the body was dispatched at one of the larger towns, where funeral facilities were more likely to be available.

Later, Pullman devised simpler cars, which were cheaper but still clean and comfortable with good service. Whatever the level of luxury, the attendants were always black and male, and all known as "George." At first, they were not paid and relied on tips to earn a living. Although this changed later, tips remained a key part of their wages. It was a tough job, according to one historian: "a cross between a concierge, bellhop, valet, housekeeper, mechanic, babysitter, and security guard." To make it tougher, Pullman sent inspectors incognito, to ensure that the attendants carried out their tasks properly. These inspectors would "mislay" jewelry, and the female ones would even make romantic overtures to tempt the attendants into breaking the rules. Attendants who failed to respond appropriately were fired instantly. Despite the many indignities, being an attendant was a stable and reasonably well-paid job, so was much sought-after.

By the early 1870s, Pullman had become, according to one railroad historian, "the foremost industrial name in the United States." He was to remain so for more than 20 years. Pullmans were introduced in Europe and Asia, too, but it was the dining-car concept that really caught on. Pullman's influence spread outside the US, though

TRAVELING IN STYLE
This 1910 poster promotes the comforts of Pullman Palace dining cars on the Chicago and Alton Railroad, the first railroad Pullman ever worked on.

railroad companies tended to provide their own sleeping cars or use those of Pullman's rival. Pullman's cars also spawned the trend for luxury that was continued by the Palace on Wheels in India, the Blue Train in Africa, and most famously on the Orient Express (see pp.190–97).

The name of Pullman also lives on in a town he built east of Chicago to house the workers at his factory. The factory is long abandoned but the pleasant, well-laid-out houses survive. The housing, however, was a source of friction between Pullman and his employees. When the economic panic of 1893 reduced demand for new cars, Pullman announced layoffs in the factory, but refused to reduce the rents on the houses he provided for them. A bitter strike ensued and spread across the country after 250,000 railworkers joined the action in sympathy. Violence broke out in several cities across the US and the strike eventually collapsed, but it left a bitter legacy. Pullman's reputation was so tarnished among workers that when he died, his family arranged for his remains to be placed in a lead-lined mahogany coffin, which was then sealed inside a block of concrete for fear that it would be dug up by angry trade unionists. It was not a comfortable end for a man who brought a good night's sleep and a decent hot meal to millions of rail passengers.

Railroads
Come of Age

TEE SAPHIR
CLASS VT 11.5
DIESEL (DMU), 1957

By the last quarter of the 19th century, the iron road had become so profitable that many railroad companies were using their profits to create new lines. It was an adventurous period that saw the construction of many extraordinary railroads, including the Trans-Siberian—a 5,750-mile (9,250-km) line between Moscow and Vladivostok, which remains the longest railroad in the world.

Railroad technology was now tried and tested, and promoters were driving it to its limits. In South America, several lines were built through the Andes to help exploit the region's mineral wealth—becoming not only the highest railroad lines in the world, but also the most spectacular, cutting through mountains and running along perilous cliffs and precipices. In India, the British desire for a cooler climate in summer led to the construction of a series of narrow-gauge hill railroads that climbed slowly but surely up the steep inclines far quicker than the road traffic ever had. One of the most ambitious ideas was to run a railroad line across the whole of Africa, from Cape Town to Cairo. Cecil Rhodes, the prime minister of Cape Colony (in what is today South Africa), had hoped to link the whole continent by a railroad that traveled through only British colonies, but was stymied by difficulties with construction, lack of finance, and the unwillingness of the British government to support the plan.

It was also a time in which services were greatly improved to encourage passengers (particularly the rich) to use the trains. The most notable luxury service was the Orient Express, which crossed the whole of Europe from Paris to Constantinople (Istanbul). In the United States, competition between railroads led to the launch of rival services between New York and Chicago, offering the red carpet treatment to their passengers. And further south, in Florida, the "overseas express" to Key West—one of the most astonishing railroads ever built—was completed by Henry Flagler. It was an exciting time for planners and passengers alike.

The Trans-Siberian Railway

W HEN PRINCESS MARIA VOLKONSKY sped across Russia from Moscow to join her husband in exile in 1827, it took 23 days before she saw the churches of Irkutsk, the capital of eastern Siberia, looming out of the snowy atmosphere. That was extremely fast by contemporary standards, and she had traveled night and day on the trans-Siberian Trakt—a crude road that was easier to travel in winter. At other times of the year, when the road was muddy, travelers could take up to nine months to reach Irkutsk, which was barely two thirds of the way to Vladivostok, the port on the Pacific that would become the terminus of the Trans-Siberian Railway.

Siberia had long been equated with exile and little else. It was a distant part of Russia, a huge region encompassing all Russia east of the Ural mountains. A spartan land, its small population was concentrated on a few river and road arteries, and was mostly employed to maintain the Trakt or guard the territory. They were supplemented by two types of exiles: criminals sent to Siberia as an alternative to prison or execution; and political exiles like Prince Sergei Volkonsky, Maria's husband, who had been involved in a failed coup attempt in December 1825.

The terrible transportation situation between Siberia and European Russia provided the impetus to build the Trans-Siberian Railway, which was by far the most ambitious railroad project ever attempted. Russia had first established a base on the Pacific as early as the 17th century, but its control of the land between the Urals and the ocean was maintained only tenuously. Indeed, Russia's ability to retain its vast Eastern territory began to look tenuous in the mid-19th century, as the development of efficient steamships in the 1840s and the completion of the Suez Canal in 1869 made it easier for its European rivals—France, Britain, and Prussia—to access the Pacific. The completion of the first American transcontinental, also in that year, followed in 1885 by its Canadian equivalent, raised fears among the Russian elite that an invasion from the East was imminent.

There had been discussions about a possible trans-Siberian line as early as in the 1850s, and a succession of plans and projects were presented to the Russian government over the following decades.

THE TRANS-SIBERIAN TRAKT
Before the Trans-Siberian Railway, the main route
through Siberia was a primitive road known as the
Trakt, seen here as it reached the outskirts of Irkutsk.

Several were madcap ideas dreamed up by foreigners intent on
making a fast buck by exploiting what they perceived to be Russian
naivety, but others were sane and realistic suggestions. Some of the
tales of these projects may have lost accuracy in the telling. A
favorite is the idea that a British gentleman going by the name of
Mr. Dull was the first to suggest a trans-Siberian railroad.
Unfortunately, the truth is more prosaic, or, rather, duller. The
individual was in fact Thomas Duff, an adventurer who went to
China and returned to St. Petersburg in 1857, where he knocked on
the door of the transportation minister, Constantine Chevkin, and
suggested the construction of a "tramway" from Nizhny Novgorod,
265 miles (426km) east of Moscow, to the Urals. The line would be
horse-drawn, and some of the four million wild horses that roamed
western Siberia could be enlisted to provide the traction.

Duff was quietly shown the door, as was a succession of both
Russian and foreign entrepreneurs. Even Nikolay Muravyov-
Amursky, the governor-general of eastern Siberia, who had managed
to establish Russia's control over previously disputed territory and
who wanted to build a line connecting the Pacific Ocean with the
Siberian interior, had no better luck than Duff. Neither did three
Englishmen (about whom little is known except that they were called

Sleigh, Horn, and Morrison), nor Peter Collins, an adventurer from New York, who was reportedly the first American to cross the entire breadth of Siberia. Collins also suggested a line in eastern Siberia, from Chita, 250 miles (400km) east of Lake Baikal, to the navigable section of the Amur River, which flows into the Pacific.

Despite rejecting all these suggestions, controversy raged within the Russian government throughout this period about the need for a trans-Siberian line. While there were many reasons not to build the line—such as the expense and the technical difficulties of creating a railway 5,750 miles (9,250km) long between Moscow and Vladivostok—the supporters of the project eventually won the argument on both military and nationalistic grounds. The military motives for the line were both defensive and offensive. It would not only allow a much quicker response to any attack on Vladivostok but, and this was not discussed openly, it would also make it easier for Russia to establish control over its vast but at the time very weak southern neighbor, China.

Consequently, in 1886, after some three decades of prevarication, the czarist government, despite being ruled by the very conservative Alexander III, took the radical step of deciding to build the line. The immediate catalyst for the decision seems to have been a fear that large numbers of Chinese were infiltrating Transbaikalia, the region around Lake Baikal. In fact, this had little basis in reality, but somehow it was the crucial development that finally made the government decide to give the go-ahead to the plan.

Inevitably, finding the money and getting the unwieldy Russian government behind the plan delayed the start of construction for a further five years. Finally, however, the czar dispatched his son, the future Nicholas II, to Vladivostok, where on May 31, 1891, he wielded a shovel to fill a wheelbarrow with clay soil, which he emptied onto an embankment of what would become the Ussuri line. Having done so, however, there was still no agreement on how to complete the work, or how it could be funded. What the project needed was a man of vision and drive to see it through to its completion—and that is just what it received. His name was Sergei Witte, and he was briefly transportation minister in the Russian government, but was finance minister by August 1892. Normally such posts were held by people of limited vision, with an interest only in keeping the purse strings tight—not so with Sergei Witte. Using his skills as a math graduate, he both managed the country's finances with acumen and ensured that there were plenty of funds for work on the Trans-Siberian Railway.

Born in the Georgian capital, Tblisi, Witte came from a minor aristocratic family that had fallen on hard times, and he had to work as a railway clerk—a very lowly task for a man of his birth and ambition. His ability was soon recognized, however, and he was swiftly promoted to manage a railroad company, and then obtained a senior government post in St. Petersburg. When he became finance minister, work on the railroad had come to a halt due to a famine in the Volga region and lack of funds. Witte's masterstroke was to create a Committee for the Siberian Railway, headed by the young czarevitch (heir to the Russian throne) Nicholas, which effectively guaranteed that the project would enjoy the continuing support of the monarch. Witte thus became the father of the railroad. He took a constant interest in its progress, ensured that money was available, fought off any resistance to the project within government, and appeased the Chinese, who were highly suspicious that the line would be used against them.

The difficulties facing the railroad's builders can hardly be exaggerated. Although the terrain the line had to cross was not as difficult as the Alps (see pp.102–107) or the Andes (see pp.198 203), nor as barren as the American deserts (see pp.32 39), the sheer length, the extreme temperatures, and the absence of a local labor force made the railroad's construction an unprecedented challenge.

To illustrate the scale of the task, the 5,750-mile (9,250-km) route was 2,000 miles (3,200km) longer than the Canadian transcontinental—and the US equivalent was not only shorter overall, it only required 1,750 miles (2,800km) of new track, since a great deal

SERGEI WITTE
Russian finance minister Sergei Witte oversaw the industrialization of much of the Russian Empire. He was the driving force behind the building of the Trans-Siberian Railway.

HARD LABOR
Workers lay tracks for the central portion of the
Trans-Siberian Railway, in the Krasnoyarsk region
of Russia. This section of the line runs between the
Ob and Yenisey rivers.

of railroad had already been laid in the east. By contrast, Russia's
railroads only reached as far as the Urals, and so the Trans-Siberian
Railway needed an entire 4,500 miles (7,240km) of new track.

Although there were no enormous mountain ranges in the way—
the Urals and the Siberian ranges were relatively easy to get through—
there was no shortage of other difficulties. In the vast steppe, neither
stone for ballast nor wood for ties could be sourced locally, so materials
had to be brought from afar, mostly by river. The rails, too, had to be
transported from factories in the Urals and eastern Russia, as did the
steel for the bridges, which had to cross the massive Siberian rivers.
Then, two-thirds of the way from Moscow, was the biggest obstacle of
all—Lake Baikal, Russia's biggest lake by volume and the deepest in the
world. The northern shore was too much of a detour and the southern
one was lined with steep cliffs right to the water's edge, which meant
that a shelf for the railroad had to be blasted out of the stone.

Time was at a premium, with the Czar intent on seeing the project completed within a decade. As a result, the surveys of the route were cursory, carried out only on a narrow belt that had been drawn thousands of miles away by St. Petersburg bureaucrats who had never been to Siberia and had only inaccurate maps. The ethos behind the construction was to "muddle through," since it was reckoned that building the perfect railroad would simply take too long. That strategy worked well in terms of ensuring that the job was done on time, but the result was a very basic railroad that could only carry a handful of trains per day and was dogged with technical problems in its early years.

For construction purposes, the railroad was divided into three main sections, each of around 1,500 miles (2,400km)—the western, the mid-Siberian, and the far eastern—and it was the latter that had the most difficulties. Work started first on the western section in 1891, from Chelyabinsk, the easternmost point of the existing railroad, and the main difficulty was a lack of local workers. It was estimated that some 80,000 men would be needed to build the first two sections, so workers were recruited not only from western Russia but as far afield as Persia, Turkey, and even Italy. The work was onerous, but well paid—agricultural workers could earn far more than they did on the fields, but even then they would often return to their villages at harvest time to help their relatives. Oddly enough, the main shortage of material on this section was wood—the local lumber was deemed unsuitable—and supplies had to be brought in from western Russia.

Construction on the mid-Siberian track began in 1893, and the labor shortage was so acute that it proved necessary to call on an obvious local source of workers—convicts who had been exiled to Siberia. This proved to be an excellent decision, for they were keen

"We must give the country such industrial perfection as has been reached by the United States of America"

SERGEI WITTE

workers, not least because a year of their sentence was remitted for every eight months they worked on the railroad, and they had access to tobacco and even occasionally alcohol in the work camps.

Conditions were harsh for the workers, but they were generally better than those of other 19th-century railroad projects—largely because workers were in short supply and their employers had to keep them happy. In the summer months, between May and August, the hours were long, with men being expected to work from 5am to 7:30pm, broken only by a lengthy lunch period of an hour and a half. In the winter, work was confined to the daylight hours, but since the line was quite far south—roughly on the same latitude as London, Berlin, and Prague—this still meant a seven- or eight-hour day in mid-winter.

The work was dangerous, too. The death rate was calculated at around two percent, which was less than on other projects such as the Panama railroad (see pp.110–19) and the never-completed Cape to Cairo railroad (see pp.214–221), but it is still shocking by today's standards. The most perilous work was constructing the major bridges, which was particularly perilous in winter, when men had to perch high above the rivers with no safety equipment and were dangerously exposed to the elements. Often men became so cold that they fell unconscious and plunged to their deaths in the icy waters.

The two western sections were completed by 1899, enabling trains to reach Irkutsk, but the eastern section proved more difficult. Witte agreed to a fateful change in plan—to run the eastern section through Manchuria, part of China, rather than building the planned Amur Railway, which would have kept it on Russian soil (the Amur line was eventually built between 1907 and 1916). The Manchurian route was shorter, but it was dangerous politically. Although the Chinese government acquiesced to the arrangement, it would prove politically troublesome and eventually lead to the Russo-Japanese war, which broke out in 1904, soon after the completion of the line.

After the opening of the Chinese Eastern Railway in November 1901, there was still one section left to be built. This was the 110-mile- (180-km) long Circum-Baikal Railway, along the southern shore of Lake Baikal, which presented the most severe difficulties. Work did not start until 1895 and, because of the need to create a shelf in the cliffs, it did not finish until 1905. Until then, passengers traveling east of Irkutsk had to take a train ferry across the lake in summer, or a sleigh over it in winter. In fact, it was not until 1916—when the Amur

Railway, which required the erection of the longest bridge on the line at Khabarovsk, was completed—that the whole journey from Moscow to Vladivostok could be undertaken entirely on Russian soil.

Even the most optimistic promoters of the railroad could not have anticipated the impact that it would have on Siberian—and indeed Russian—history, and it has not all been good. Not only was the line the catalyst for the Russo-Japanese war, but it also played a key role in several other conflicts, most notably two world wars. Also, the czarist regime that created it paid a heavy price. By concentrating so many of its limited resources on the project, it neglected other areas of spending, and this imbalance helped trigger the revolution that led to the overthrow of the monarchy in 1917. This also led to the execution of Nicholas II and his family at Ekaterinburg, which, ironically, is one of the main stations on the western section of the line. Nevertheless, the project must be counted as a success, despite its cost, and the sometimes unusual conditions endured by its early passengers (see p.146). The Trans-Siberian remains the main artery between Siberia and the rest of Russia. It is a double-track, electrified line and is heavily used by both freight and passenger trains. It is the longest railroad in the world, and arguably the most important.

RUSSIAN STEAM LOCOMOTIVE
The USSR stopped producing steam engines in the 1950s, but a few are preserved at stations on the Trans-Siberian Railway. This model remained in service until the 1970s.

RUSSIAN TIME ZONES

Such is the length of the Trans-Siberian Railway that the journey from Moscow to Vladivostok crosses seven time zones. Timetables and station clocks are set to Moscow time (MT), which results in discrepancies between local and railroad time. These discrepancies increase from two to seven hours the further one travels east.

KEY

- ■ MT
- ■ +2
- ■ +3
- ■ +4
- ■ +5
- ■ +6
- ■ +7

MOSCOW

VLADIVOSTOK

ST. PETERSBURG

Lake Ladoga

KARA SEA

VOLOGDA

MOSCOW

NIZHNII-NOVGOROD

RAZAN'

Kirov

URAL MOUNTAINS

PERM

SYZRAN

SAMARA

UFA

EKATERINBURG

Tiumen

Zlatousr

Mass

Cheliabinsk

Kurgan

OMSK

Ob'

Tomsk

Petropavlovsk

Novosibirsk

KAZAKHSTAN

SIBERIAN RAIL BRIDGES

Hundreds of bridges were built along the route of the Trans-Siberian Railway. Many of these are of unsual design, such as this lenticular bridge, otherwise known as a "fish-bellied truss". The longest bridge of all, which crosses the Amur River at Khabarovsk, is 8,570ft (2,612m) in length.

The Trans-Siberian Railway

Russia's railroad network is exceeded in size only by those of the US and China, so it is perhaps fitting that it features the world's longest continuous railway line. At around 5,750 miles (9,250km) in length, the Trans-Siberian Railway straddles the Russian interior, linking the country's heartland in the European west to its hinterland in the Asian east. A broad-gauge, double-track line that was fully electrified in 2002, it is a vital artery that has spread industry and commerce across this vast territory.

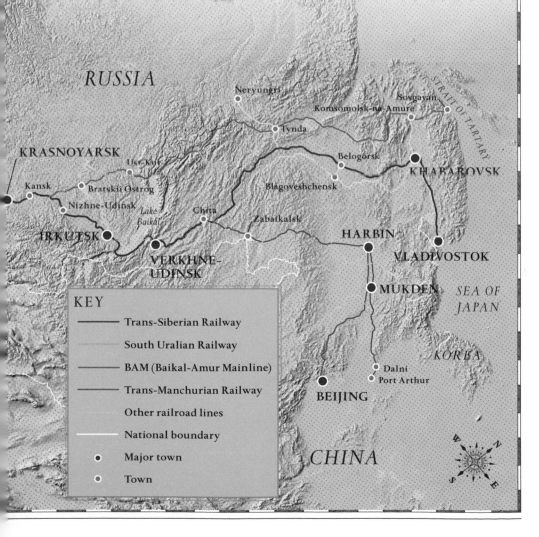

KEY

— Trans-Siberian Railway

— South Uralian Railway

— BAM (Baikal-Amur Mainline)

— Trans-Manchurian Railway

— Other railroad lines

— National boundary

● Major town

○ Town

The Orient Express

DISAPPOINTINGLY FOR SOME, there never was a murder on the Orient Express, but there is no doubt that Agatha Christie chose a fitting setting for her "whodunnit" mystery novel. The Orient Express was the most exciting and exotic train service in the world, crossing the whole of Europe and entering territories—particularly in the east—that were little known to western Europeans. Indeed, the service was one of the wonders of the age, and like so many railroad innovations it owed its existence to the tireless efforts of one individual, in this case Belgian engineer Georges Nagelmackers.

Nagelmackers was the founder of the Compagnie Internationale des Wagons-Lits, a popular service that offered compartments instead of the open-plan cars of Pullman's sleeping cars (see pp.170–77). However, Nagelmackers' real genius lay not in the trains he built but in the routes he established. He wanted a Europe *sans frontières*, one that travelers could cross quickly and in style in his well-appointed trains. To that end, in 1872, he created a service that ran from Ostend on the North Sea coast of his native Belgium more than 1,000 miles (1,600km) south to Brindisi on the tip of the Italian heel. The venture proved successful, and with the East and the Balkans opening up as the Ottoman Empire declined, he saw that a service linking Europe and Asia would also be profitable. And so he began work on his Orient Express—an 1,857-mile (2,989-km) passage from Paris to Constantinople (now Istanbul), bridging east and west and crossing six countries en route.

Dealing with the railroads of six diverse nations was no easy task, and Nagelmackers had to use all his skills as a negotiator to solve a whole range of problems. Most importantly, he had to ensure that each

GEORGES NAGELMACKERS
Belgian industrialist Georges Nagelmackers was the founder of the Compagnie Internationale des Wagons-Lits, the company best known for its Orient Express service.

THE ORIENT EXPRESS

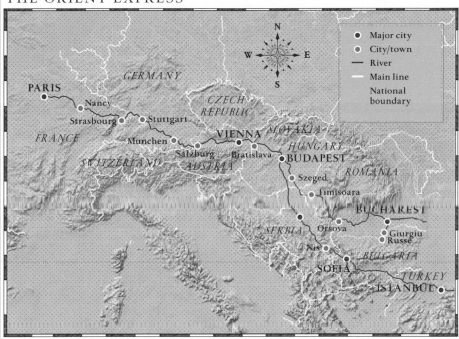

country had locomotives that could haul his trains, and that the tracks were of standard gauge (see pp.90–91). Other issues included the width of the route's tunnels, and arcane matters such as the security of wine lockers. He was also a great publicist and generated huge interest in his venture, not least because the route passed through the Balkans, an area that was still recovering from numerous wars, having struggled for independence from both the Ottoman and Austro-Hungarian empires.

The inaugural train, for press and VIPs, left the Gare de L'Est in Paris on the evening of October 4, 1883, and was scheduled to take three and a half days to reach Constantinople. Nagelmackers had created what *The Times* correspondent Henri Opper de Blowitz described as a level "of comfort and facility hitherto unknown." There was a smoking room, a ladies' boudoir, and a library, and each compartment (or coupé) had a miniature drawing room in the style of Louis XIV, complete with a Turkish carpet, inlaid tables, and plush red armchairs. In the evenings, the compartment walls could be folded down to reveal beautifully upholstered beds. The *cabinet de toilette* had a mosaic

ALPINE ADVENTURE
The Orient Express passes through the Austrian Alps shortly
before World War I. Here it is hauled by an Austrian locomotive—
one of six different engines that would take it to Constantinople.

floor, and a special car at the rear of the train had cubicles for showers
that were supplied with hot and cold water—certainly a railroad
first. According to one account (and numerous tales were published)
the *piéce de résistance* was the dining room:

> [It] had a ceiling with embossed leather from Cordoue [Cordoba], walls
> lined with tapestries from the Atelier des Gobelins, founded by the Sun
> King, and drapes of finest Gènes [Genoa] velvet.

The tables were covered with white damask cloths and intricately
folded napkins, and ice buckets filled with champagne bottles were at
hand—and if the five-course meal were not enough, iceboxes full of
exotic foods and cold drinks were available at the end of each car.

Nagelmackers was a stickler for detail, and set out a series of rules
to maintain high standards. Attendants had to be smart at all times
and on special occasions had to dress like footmen from the time of
Louis XIV, complete with blue silk breeches and buckled shoes. Even
the engine crew had to dress up on occasion, often in white coats
that were highly impractical in the engineer's cab.

That first journey was one long party. At Strasbourg, Vienna,
and Budapest the train was met by brass bands and local dignitaries,
while at Tsigany, in Hungary, a gypsy orchestra came aboard and

serenaded the passengers all the way to the border with Romania. The only drawback was that the track was incomplete. The bridge over the Danube river was unfinished, so the crossing from Romania to Bulgaria had to be made by ferry, and even then the line only reached as far as the port of Varna, where passengers had to take a ship to Constantinople. This last section of the journey was, according to Blowitz, through a land full of "brigands" who had recently attacked one station and "garrotted the stationmaster and his subordinates in order to get hold of the money they expected in his till," and only fled when they were disturbed by workmen. Consequently, Blowitz and his companions armed themselves with revolvers, although they never had occasion to use them. They arrived at Constantinople precisely 82 hours after their departure from Paris and were met by the Sultan, with whom Blowitz conducted a newspaper interview, the ruler's first.

It was another six years before it was possible to take the train all the way to Constantinople. The trip took just under three days, leaving Paris on a Wednesday at 7:30pm and arriving at 5:35pm on the Saturday in Constantinople. The service became popular, attracting a wide range of travelers as it was quicker and more convenient than traveling by ship. Subsequently, a variety of other routes were opened, running trains that bore some variant of the name "Orient Express." Various connecting trains were also introduced, including one from London via a train ferry. Inevitably, when war broke out in 1914, the Orient Express was suspended, but as soon as the war was over in 1918 a second line was opened—the Simplon Orient Express, on which Agatha Christie's tale is set. Using the Simplon Tunnel between Switzerland and Italy, this second line was a more southerly route via Milan, Venice, and Trieste, and soon became the more popular route from Paris to Constantinople. A third line was

BOUND FOR PARIS
On April 18, 1887, a traveler bought this ticket for Fr52.95. It afforded passage from Budapest to Paris aboard the Orient Express.

LUXURY DINING
The dining car of the Orient Express, set for dinner. The very first menu included oysters, turbot with green sauce, chaud-froid of game animals, chicken à la chasseur, and a buffet of desserts.

added in the 1930s (the heyday of the Orient services)—the Arlberg Orient Express, which ran via Zurich and Innsbruck to Budapest, with sleeper cars continuing on to Bucharest and Athens. The outbreak of World War II in 1939 again interrupted the service, although the German Mitropa company ran trains through the Balkans until Yugoslav partisans destroyed the line.

While there were no recorded murders on the Orient Express, there was at least one mysterious death—when a US agent fell from a train at the height of the Cold War in the 1950s—and there was no shortage of mischief. Sleeping car attendants were regularly called on to hire prostitutes, not just for gentlemen, but for princes and even bishops who found that a train offered more privacy than a brothel. Indeed, many men took the train simply for these services alone. There was plenty of spying, too, given that the trains linked east and west. Between the wars, "King's Messengers"—effectively couriers for the British Foreign Office—traveled in compartments where they guarded their diplomatic bags with their lives—and later claimed that they were immune to the wiles of the beautiful young spies who were sent to entrap them, although we only have their word for that.

For all its luxuries, however, the service did not remain exclusive for long. On the contrary, although there was only one class of car in the original service, which cost 300 Francs (the equivalent of two weeks' average wages at the time), second and third class cars were opened for poorer people

EASTERN PROMISE
A French poster displays the timetable of the Orient Express over the winter of 1888. Its seven main stops are advertised, plus a connecting service to London.

"Peasants in half a dozen countries would pause in their work in the fields and gape at the glittering cars and the supercilious faces behind the windows"

E.H. COOKRIDGE, *THE ORIENT EXPRESS*

who used it for shorter, mostly domestic journeys. In these packed cars, as one writer put it, "the pulse of Old Europe beat, with its almost medieval characters: the tramp, the peddler, the gypsy-musician..." Indeed, although the service continued to operate even after the Iron Curtain divided Europe, the communist countries increasingly replaced the luxurious Wagon-Lits cars with more spartan cars run by their own railroad networks.

By 1962, the Orient Express and the Arlberg Orient Express had stopped running, leaving only the Simplon Orient Express, which was replaced by a slower service, the Direct Orient Express, which ran daily trains from Paris to Belgrade and twice-weekly services from Paris to Istanbul and Athens. The service through to Istanbul finally came to an end in 1977, killed off by the ubiquitous spread of the automobile. A service called the Orient Express stuttered on between Paris and Vienna, and between Budapest and Bucharest, but that closed with the opening of the high-speed line between Paris and Strasbourg in 2009 (a service called the Venice-Simplon Orient Express still operates between London and Venice, but it is a separate entity). By the time of its swansong in 2009, the original Orient Express was an anachronism—a relic that had perhaps done well to last as long as it did.

The Most Spectacular Railroads in the World

HENRY MEIGGS WAS A SCOUNDREL who made good building railroads. When he arrived in Chile in 1855, he was a 44-year-old outcast who had fled San Francisco amid accusations of fraud, but by the time he died 22 years later, this handsome, larger-than-life character had been honored with the unofficial title of "Don Enrique." By then, he had conquered some of the world's most difficult railroad terrain—the seemingly impossible slopes of the Andes—for which he also earned the nickname "Yankee Pizarro," after the Spanish conqueror of the Incas.

In his youth, Meiggs demonstrated a remarkable capacity for hatching imaginative schemes, but he was never entirely honest—if a venture failed, he was not above lying and manipulating others to avoid being exposed. He had an early success in the lumber business, setting up his own company in New York City and relocating to San Francisco during the Gold Rush. After that he went into property and developed land near the Golden Gate, but he soon fell into debt and only avoided bankruptcy by raising cash with illicitly obtained warrants. When his fraud was discovered, he fled to Chile, where his devious reputation preceded him. The only work he could find was overseeing a gang of laborers building the railroads, but he proved to be so efficient that he was soon given charge of entire railroad projects. Two of his predecessors, a New Englander named William Wheelwright and the great Richard Trevithick (see pp.20–21), had dreamed of building lines inland from the west coast of South America, but it was Meiggs who finally built them. This achievement had enormous consequences for the region, for it opened up vast deposits of copper, silver, and minerals for exploitation, and made fortunes for Chile and Peru.

HENRY MEIGGS
Charismatic American entrepreneur Henry Meiggs pioneered the railroad lines of Chile and Peru. Doing so required scaling the heights of the Andes.

To help him with construction, Meiggs engaged engineers who shared his own daring outlook on the fearsome terrain. His first big success came with the laying of a 90-mile (145-km) line from the Pacific coast to San Fernando, which involved bridging the Maipo River, previously a major obstacle separating north and south Chile. On completing the line faster than his contract required, Meiggs successfully bid for the most important route in the country—from Valparaiso, on the west coast, to Santiago (the capital), a mere 55 miles (89km) inland as the crow flies, but in practice 115 miles (185km) due to the mountainous terrain. Wheelwright had begun the route, but after spending a million pesos (or tens of thousands of dollars) only 4 miles (6.5km) had been completed by the time Meiggs took over. To finish the route, the government borrowed money from Barings Bank of London, and Meiggs did a speedy deal with the Minister of the Interior, promising to complete the route in three years for six million pesos—so long as he received an additional half million pesos if he finished early, plus an extra 10,000 pesos for every month gained. Backed by a workforce of 10,000 men, Meiggs completed the line in just two years and three days, a triumph that proved his astonishing capabilities both as a contractor and a negotiator.

Having thus "conquered" Chile, Meiggs moved on to even bigger projects in Peru, a country that was just striking it rich thanks to its enormous deposits of guano, or bird droppings, which made an excellent fertilizer. Understandably, Peruvians wished to use their new-found wealth to build a railroad system that would unify the country, just as the Belgians had done forty years earlier (see pp.43–44), and as the Canadians (see pp.125–27) and the Italians (see pp.46–47) were doing at the time. For that reason Meiggs was welcomed with open arms—or open palms, in the case of the ruling class, who demanded huge amounts of money in bribes.

Meiggs's big opportunity came as the result of an episode that was typical of Peru's dramatic political history. In 1868, Colonel José Balta, the type of buccaneering officer often found in South America at the time, was elected president. Immediately after his election, Peru suffered a devastating earthquake, and Meiggs cannily donated $50,000 to the government, or rather to Balta personally, ostensibly for crisis relief. Balta had already upset the local oligarchy by giving a French company a monopoly on the sale of guano, and now he used money raised by the deal to pay another foreigner—Meiggs—to build Peru's railroads. And so, in the three years following Balta's

HEIGHT ABOVE SEA
LEVEL OF GALERA
STATION IN PERU

15,681FT

(4,777M)

THE WORLD'S
HIGHEST STATION

election, Meiggs signed six contracts to build over 1,000 miles (1,600km) of railroads, on terms that were highly favorable to him. The result was that, having had a mere 61 miles (98km) of track in 1861, Peru had 947 miles (1,524km) by 1874, and nearly 2,000 miles (13,200km) by 1879, two years after Meiggs' death.

Meiggs built two lines in Peru, and both are wonders of the railroad world. The first runs from the southern port of Mollendo to Arequipa, Peru's second-biggest city, and then up to Lake Titicaca and the mining area of Juliaca. Meiggs estimated that it would cost him 10 million soles to build it (around $300 million today), and then told the government that it would cost 15 million, and proceded to build it for 12 million (and even completed it early). The second line, the Central Peruvian, rises from Callao, the port of Lima, up through the steepest and highest sections of the Andes, following precipitous llama paths to the copper mines of Huancayo and the fabled silver mines of Cerro de Pasco.

Unfortunately, the lines were built at a time of great financial turmoil. As the country's guano ran out, so the supply of money from the government dried up, and Meiggs was forced to use his own bills of exchange—the so-called "Billetes de Meiggs." Sadly, too, Meiggs died during construction, but by then he had shown how it was done, and had conquered the steepest slopes. He followed the British idea of zig-zagging railroads uphill (see pp.204–205), but did so on an unprecedented scale. In India, the British lines reached 2,500ft (760m)—Meiggs's trains scaled mountains over 14,000ft (4,250m) high, and zig-zagged 25 times in a matter of 100 miles (160km).

Being a foreigner, Meiggs was an easy scapegoat for Peru's economic woes, one journalist even writing that the "the ruin of Peru is the monument [to] Henry Meiggs." However, he was mostly considered a hero, and one of the country's highest peaks was soon named after

BRIDGING THE ABYSS
To access the remoter parts of Peru, Henry Meiggs constructed numerous bridges, such as this seemingly precarious viaduct in the mining region of Juliaca.

BILLETES DE MEIGGS
In the 1870s, Henry Meiggs effectively created his
own currency to finance his railroad projects in
Peru. Around a million soles' worth of banknotes
known as Billetes de Meiggs were put in circulation.

him. His success was due not only to his ability to find the best route
for a railroad, but also to his formidable organizational skills and his
ability to bring out the best in his workers. One Peruvian journalist
described his "railroad army" battling the elements:

> The 'army' (distributed along the line in eleven camps), consisting of
> Don Enrique's engineers and labourers, was attacking the Andes. The
> scouts went ahead to determine the best and least costly route; the
> advance guard followed in their tracks, staking out the exact route to be
> followed; next came the main body, levelling the barriers, making fills
> and cuts and piercing tunnels; lastly, there was the rear guard, putting
> down ties and laying rails.

Meiggs was famously generous to his workforce, particularly the
rotos—the much-feared, much-despised, Chilean working class.
According to James Fawcett, in *Railways of the Andes*, a typical *roto* was
notorious: "for his hardihood, his skill in the handling of the sharp,
curved, disembowelling knife that all his tribe carried, his hatred of
any sort of discipline, his love of cane sugar as a beverage and his
addiction to gambling." Meiggs succeeded by treating his workers as
men rather than slaves, and he was even more successful with the
5,000 Chinese workers he hired and who were normally treated
worse than the *rotos*. (see pp.88–89). According to one observer,

quoted by Fawcett: "some of them were fat, the only fat Chinese in the country! Meiggs fed them well with rice and beef in plenty and a good breakfast of bread and tea before starting the day's work."

Toward the end of his life, Meiggs wanted to return to the US, claiming that he had repaid his San Francisco debts, but the governor of San Francisco vetoed a bill that was passed to exonerate him from his offences. In 1977, a century after Meiggs's death, the California Supreme Court quashed the indictment against him, declaring that he "had gone to a higher court," but his death was not the end of his family's influence. His nephew, Minor C. Keith, went on to complete a railroad his uncle had started in Costa Rica. Its major source of income was carrying bananas, and Keith went on to found United Fruit, the colossus that dominated the sale of bananas for a century.

Going Uphill

The first railroad lines were laid along the flattest routes possible, but the requirements of industry—particularly mining—soon demanded that trains could travel uphill. The earliest solution was the "incline"—a pair of parallel tracks that enabled a descending train to haul an ascending train uphill by means of a chain that connected the two via a pulley. This worked particularly well on the short sections of track that were used to draw raw materials up from pits and quarries, and variants were powered by horses or stationary steam engines. The basic premise lies behind modern-day funicular railroads, and other engineering solutions, such as spiral loops and switchbacks, have also been developed.

Spiral loop

Building a railroad track on a spiral allows a train to gain elevation in a much shorter length of track than would be possible with a conventional curve. Spiral loops also avoid the inconveniences of reverse travel and interrupted movement that are necessary when climbing switchbacks, the other railroad engineering method by which traction trains climb hills (see right). Popular in challenging terrain, such as mountainous regions in which level ground is limited, spirals are set at a constant grade and degree of curvature, and allow the track to pass over itself as the line ascends.

THE BRUSIO SPIRAL VIADUCT IN SWITZERLAND WAS BUILT IN 1908. IT DESCENDS 60FT (20M) AT A GRADE OF 7 PERCENT.

SWITCHBACKS
The railroad linking Ecuador's coast with its capital, Quito—at 9,350 ft (2,850m)—crosses the Nariz del Diablo escarpment via a series of switchbacks. By this method, trains gain height in a short length of track by entering a dead-end siding, and then reversing to climb the next switchback.

Funicular railroads

The first funicular railroad—a form of cable car operating on similar principles to a railroad incline—opened in 1862 in Lyon, France, and featured a four-rail track layout on which two cars traveled on separate parallel tracks. The development of sidings allowed later designs to economize on the space and materials used. First came a three-rail layout, in which cars shared a central rail—then came a two-rail layout in which the cars shared both rails on either side of the siding.

HOW IT WORKS

Two- and three-rail funiculars contain a siding at the halfway point. Each car has "blind" (flangeless) inner wheels and double-flanged outer wheels to prevent them from switching rails.

Descending car

Two-rail funicular

Weight of upper car hauls lower car up the slope

Siding

Pulley cable

Flangeless wheel rides on top of the inside rail

Outer wheel with flanges on either side of outside rail

THE KIEV FUNICULAR, UKRAINE, OPENED IN 1905. IT CLIMBS 781FT (238M) OF TRACK AT A GRADE OF 36 PERCENT.

Henry Flagler and the Overseas Railroad

HE WAS UNIQUE in the history of the United States. He could have been as rich and as famous as John D. Rockefeller. Instead, Henry Flagler chose to spend the last 30 years of his life building a railroad line in Florida and establishing Florida's tourist industry. His final triumph, completed only a year before he died in 1913, was the construction of the world's most ambitious "overseas railroad"—a line stretching all the way from the mainland across the Florida Keys to the southernmost tip of the US, Key West.

For decades before his great Florida adventure, Flagler had been a key partner in the creation of Rockefeller's gigantic Standard Oil monopoly. According to Flagler's biographer, David Chandler, Rockefeller readily admitted that Flagler was an inspiration. Indeed, Flagler had contributed more than Rockefeller to the organization of the Standard, and was wholly responsible for the clever legal structure that protected it against antitrust lawsuits. However, Flagler craved an outlet for his colossal creative energies, and that was just what he found in Florida, which he explored in 1883 while on honeyoon with his second wife, Ida.

At the time, Florida was still a young state and was keen to sell its land rights, and Flagler saw that even underdeveloped St. Augustine, the oldest European settlement in the US, was attracting plenty of wealthy tourists. Sensing a golden opportunity, he gave up his daily involvement in the Standard, and set about building a chain of hotels along the east coast—the first of which, the 540-room Ponce de León, opened in St. Augustine in 1888.

Designed by the architects of the Metropolitan Opera and New York Public Library, the Ponce de León was the height of luxury and immediately attracted visitors. However, the local railroads were a deterrent, for they ran on a variety of gauges (see pp.90–91) and so demanded frequent changes of train. What was needed was a reliable railroad that would link the town directly to New York, so Flagler bought up the existing lines and converted them to standard gauge. As he noted: "the average passenger will take a through car ninety five times in a hundred in preference to making a change".

FLORIDA'S EAST COAST RAILROAD

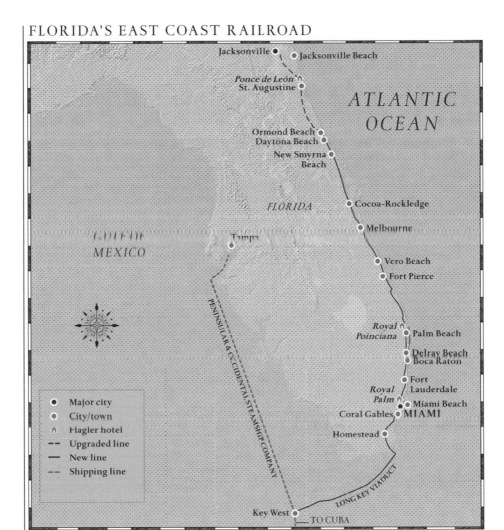

Another problem was that the lines only ran to Daytona, a beach about a third of the way down the coast—beyond that, Flagler would have to lay his own tracks.

Flager's next port of call was Palm Beach, a natural harbor that he explored in 1893. As always, he inspected the site incognito, to avoid attracting attention, and then returned openly to buy the land he wanted. Within months he opened the 1,100-room Royal Poinciana hotel, which Chandler called "the largest resort hotel in the world... equipped and staffed in the most luxurious manner imaginable," and extended the railroad south to reach it. The guests did full justice to

PONCE DE LEÓN HOTEL
Completed in 1888, the Ponce de León hotel was a landmark in Henry Flagler's development of Florida's east coast. Today it lies at the heart of Flagler College, founded in 1968.

the luxury of the accommodation, with a hundred private railroad cars arriving each winter for the George Washington Birthday Ball—an event at which the most powerful men in America, including Flagler himself, dressed in the most elaborate drag costumes, complete with fishnet stockings, powdered wigs, and strings of diamonds. Another great draw was an annex called the Breakers, which proved so popular that Flagler turned it into a casino.

To foster local industry, Flagler also established a "Model Land Company," which, in Chandler's words, "did more perhaps to actually building up the Florida East Coast than any of his other undertakings." He encouraged people to plant vegetables, citrus fruit, and pineapples, and when an unprecedented snowstorm wiped out the fledgling industry in the winter of 1884 he secretly spent a fortune helping the affected farmers. He even arranged a link-up with weather forecasters, and if a serious drop in temperature was predicted, his engine drivers sounded six long blasts on their whistles as they thundered through the orange groves, calling on farmers to hurry out with their "smudge pots"—oil-burning heaters that prevented frost from forming on the fruit trees.

Another consequence of these freezes was Flagler's decision to extend his railroad south, for only 60 miles (100km) away the climate was warmer and better suited to farming. Extending the railroad also entitled him to land grants, and on the basis of these he accumulated over two million acres of territory, including Biscayne Bay, a beautiful spot protected from the Atlantic by a barrier island and watered by the little Miami River. At the time, the area was "mostly swamp, and was filled with mosquitoes, snakes, mangrove thickets and Spanish Bayonet" (a nasty type of cactus), but Flagler tamed it and brought water and electricity to the town he established there. The locals wanted to call the town "Flagler," but Flagler declined the honor, preferring to name it after the river. Thus Miami was born. The following year he set about building another luxury hotel, the Royal Palm, which was as popular as the Royal Poinciana. The railroad arrived at Miami in 1896, completing a 500-mile (800-km) line that ran south from the neighboring state of Georgia through Jacksonville and the resorts Flagler had built. Anyone glancing at a map could see that Miami was the end of the line, but Flagler kept on going, out to sea, away from the mainland and over the Florida Keys—an archipelago that sweeps southwest from Miami to Key West, far out in the Gulf of Mexico. "There is an impelling force within me," he told a friend, "and I must carry out my plans"—and the result was his 128-mile (206-km) "overseas railway."

In one respect Flagler was lucky. By the time his plans were ready, President Theodore Roosevelt had authorized the construction of the Panama Canal, making Key West a potentially vital transportation hub. As for getting the work done, Flagler simply found the right man for the job and left him to it, ignoring the question of cost, even though there were no more grants to be had. His chosen engineer was Joseph Carroll Meredith, who had already built the massive docks at Tampico on the Gulf of Mexico. However, before he got started on the offshore part of the route, Meredith had to lay 91 miles (146km) of railroad through the Everglades, an ordeal in which every kind of danger and annoyance was encountered, from apparently bottomless swamps and uncharted lakes to snakes, alligators, mosquitoes, and obstacles that only the biggest

"But that any man could have the genius to see of what this wilderness was capable and then have the nerve to build a railroad here…"

GEORGE W. PERKINS OF J.P. MORGAN, ON FLAGLER'S ACHIEVEMENTS

dredgers in the country could move. Then he faced the 37 miles (60km) of the "overseas" railroad itself, a route demanding 17 miles (27km) of bridges and 20 miles (32km) of embankments—a feat of engineering that has never been equaled. The Long Key viaduct alone measured 2½ miles (4km), and was second only to the 7-mile (11-km) Knights Key viaduct, which rested on 366 concrete columns and had a swing bridge to allow ships to pass through it. Without dry land for accommodation, the 4,000 workers were housed on enormous barges, which contained all the facilities needed for survival, including vast quantities of fresh water. However, there was still a high casualty rate, both from accidents and disease, and a hospital was built in Miami to treat the afflicted. The elements also caused havoc, not least in 1906, when a hurricane sank an accommodation barge, causing at least at least 70 deaths, and delaying construction for a year.

The whole Florida East Coast Railway project was completed at a cost of $20 million ($500 million in today's money) and in less than seven years. This was largely thanks to the loyalty of Flagler's men, one of whom told a reporter "there isn't one of us who wouldn't give a year of his life to have Mr. Flagler see the work completed." And see

it he did, opening the line to the public on January 22, 1912, tearfully exclaiming: "my dream is fulfilled, now I can die happy." He had never expected to see the project completed, predicting 20 years earlier that it would take 30 years to finish, and accurately forecasting: "I have only 20 more years to live." The line's completion was celebrated by the introduction of the Havana Express, a regular through service that arrived at Key West only 52 hours after leaving New York, giving passengers the luxury of strolling across the quay to take a ship to Cuba, a mere 90 miles (45km) away. Flagler, his life's work accomplished, died a happy man the following year.

Sadly, however, the line never prospered. It failed to attract many passengers, and proved to be unreliable since it was affected by bad weather. Consequently, the railroad went bankrupt in 1932, and the offshore track was destroyed on Labor Day 1935, during the worst storm of the century. Nevertheless, Meredith had built his railroad well. Highway 1, which replaced it, was constructed on the roadbed that Flagler had financed. Indeed, Florida, the Sunshine State, has much to thank him for. When he started work it was one of the poorest states in the Union—today it has one of the strongest economies in the world.

OVERSEAS EXPRESS
An express train thunders over the Long Key Viaduct—a 2½-mile (4-km) stretch of arches that links two islands of the Florida Keys. Today, the structure supports the Overseas Highway, or Highway 1, which follows Flagler's railroad route.

Hauling Freight

The workhorses of the railroads, freight locomotives require high power output rather than speed. The use of containers to convey goods has increased demand for freight trains, and rail is still the preferred method for moving bulk cargo, such as coal, grains, and liquids.

UNION PACIFIC NO. 25408 CABOOSE (1959)
Common on US and Canadian railroads, cabooses were the last car on a freight train, and housed some of the crew members. Thought to be named after the Dutch *kabuis* ("ship's galley"), their use declined as automatic signaling reduced the need for large crews. The pictured model has a cupola for watching the train's cargo.

B&O NO. 3684 (1966)
No. 3684 is a General Motors GP40-class diesel-electric, a versatile locomotive that could be used for "manifest" freight—hauling a range of different units in a single train—as well as heavy loads. It was the first engine capable of a power output of 3,000bhp (2,240kw) to be used by the Baltimore & Ohio Railroad.

Front-facing cab

Mounted on a 55-ft (16.8-m) frame

B&O NO. 7402 (1964)
One of 24 Baltimore and Ohio SD35-class diesel-electric locomotives, No. 7402 was powered by a 16-cylinder engine that yielded a power output of 2,500bhp (1,900kw). Its two three-axle trucks are typical of low-speed, high-weight freight trains.

NORFOLK AND WESTERN NO.522 (1962)
No.522 was the first of a series of the GP30 class from General Motors to be put into service by the Norfolk and Western Railway. It achieved a power output of 2,250bhp (1,680kw) through its diesel-electric engine.

SANTA FE SOUTHERN NO.92 (1953)
A diesel-electric "switcher" used to switch trains at rail yards, Santa Fe Southern's No.92 is a GP7-class built by General Motors. It was one of the first locomotives to use the "hood unit" design, in which a narrow body is surrounded by external walkways.

NORFOLK AND WESTERN NO.1776 (1970)
With a power output of 3,600bhp (2,680kw), No.1776 was one of 115 SD45 diesel-electrics produced by General Motors for the Norfolk and Western Railway. It was painted in a stars-and-stripes livery to mark the 1976 bicentenary of the US Declaration of Independence.

Three 48-in (121-cm) radiator fans

Locomotive weighs 139 tons (141 tonnes)

Geared to travel at 65mph (104kph)

External walkways

SBB CARGO TRAXX F140 AC (2003)
Built by Bombardier, the TRAXX class is a modular family of trains that can be adapted for a range of functions, and is the most economically successful train ever produced. The pictured model is of an electric freight train operated by SBB Cargo of Switzerland.

Cape to Cairo:
the Railroad that Never Was

IT WAS THE MOST ABSURD, the most ambitious, and the most improbable of all railroad dreams, and it failed—but only just. No continuous line from Cairo to the Cape of Good Hope—which would have linked Britain's African colonies (colored pink on global political maps of the era)—was ever built. Nevertheless, construction started on a through-rail route in the 1880s, and one did eventually emerge 40 years later, although hundreds of miles of transfers via lakes and rivers were necessary to complete the journey. However, even the project's most optimistic protagonists had accepted that maritime interruptions would be necessary. Indeed Cecil Rhodes, the godfather of the idea, stated that the project was never for a railroad that depended on traffic all the way through, but one that would "pick up trade all the way along the route," and, crucially, would run entirely through British Imperial territory.

The idea represented a microcosm of the various impulses behind the British Empire in the late-Victorian era. It combined private megalomania, commercial and financial greed, and military necessity, although it had relatively little support—especially not of the financial kind—from London. Whether the project is viewed as a failure or as a partial success, the motley band of imperialists, contractors, and engineers who worked on it built thousands of miles of railroad that are still vital for the African continent today. The route incorporated separate lines to both the Atlantic and Indian oceans and, as a by-product, created a number of new towns and cities. Lusaka, now the capital of Zambia, was described in railroad

CECIL RHODES
British imperialist Cecil Rhodes was the man behind the dream of a railroad from the Cape all the way across the "red line" of British colonies in Africa to Cairo.

historian George Tabor's *The Cape to Cairo Railway* as no more than a "lion-infested siding," and Gaberone, the future capital of Botswana, was a "remote watering hole on the edge of the Kalahari desert."

GEORGE PAULING
Larger-than-life engineer George Pauling worked with astonishing stamina and speed to build Rhodes' railroads across southern Africa.

The idea of an all-British railroad through Africa was first suggested in 1876 by the explorer H.M. Stanley (of "Dr. Livingstone, I presume" fame), in a letter to the *Daily Telegraph* newspaper. As Stanley later put it, "the railroad is the answer to Africa's pressing problems. It is the only answer to the wagon trails, decimated by rinderpest and the tsetse fly; and the one way to defeat slavery by opening up the continent to commerce and communication." Stanley's idea became a reality through the unbridled ambition of Cecil Rhodes, an imperial and political megalomaniac who made his fortune from the lucrative diamond trade of southern Africa. Rhodes was supported throughout the project by Charles (later Sir Charles) Metcalfe, an unusual blend of aristocrat and consulting engineer, who had been a friend since the pair were students at Oxford University.

To carry out his ambitions, Rhodes required someone with the engineering skills to build a railroad over thousands of miles of virtually impassable country. He was lucky to find the right man in British railroad engineer George Pauling, head of family-owned engineering contractor Pauling & Co, which already had a decade of experience building railroads overseas when it was established under that name in 1894. The firm also included George's brother Harry, four cousins—Harold, Henry, Willie, and Percy—and his brother-in-law Alfred Lawley.

In every respect, George was a larger-than-life character. He was a very big man who professed that he was "never able to reduce my weight below 16 stone [220lbs]." This was perhaps not surprising given his enormous appetite—on one occasion he consumed 300 bottles of German beer with two friends while stuck for 48 hours on a railroad line; on another he ate a thousand oysters ("small but of delicate flavor") in one sitting. As his banker Emile d'Erlanger put it (Pauling relied on the d'Erlangers' financial backing, as Rhodes depended on the Rothschilds'), he was "endowed with a physique that made light of any feat of strength and enabled him to defy fatigue or illness."

His stamina was combined with a readiness to build rough-and-ready lines, leaving bridges to wait until later. He knew that his work would prove to be durable, and it was this confidence that ensured his promises of apparently impossible speeds of construction would be kept. His financial success came from his capacity to identify almost at a glance the shortest and cheapest route for a line. Although he based his estimate on an initial outside survey and charged a fixed fee per mile, he profited greatly from his ability to find shortcuts.

The first railroads had opened in the Cape Colony, at Africa's southern tip, in 1863. To save money, they were built to a narrow gauge of 3ft 6in (1,076mm), which became known as "Cape Gauge." Development of the lines was limited by financial constraints, and the narrow gauge resulted in speeds that never averaged more than 35mph (56kph). It was the 1872 diamond rush at Kimberley, 600 miles (965km) to the north, that sparked more ambitious railroad-building

CONSTRUCTION OF THE CAPE TO CAIRO
Work started on southern Africa's first railroad at Cape Town
in 1859. By the time the tracks reached the Congo River in 1918,
the dream of a trans-African railroad had been abandoned.

plans, adding a solid financial rationale to the young Rhodes' imperial ambitions.

The rails reached Kimberley by 1885 while Rhodes, still only in his thirties, was busy building up a diamond monopoly through his company, De Beers. This first stage of the line was no easy feat, with a climb of more than 3,500ft

DISTANCE FROM CAPE TO CAIRO AS THE CROW FLIES

4,200 MILES

(6,750 KM)

(1,000m) to the dry, dusty uplands of the Karoo. Pauling ensured that the pace of construction was far faster than the eight years it had taken to lay the first 400 miles (650km) from Cape Town to Worcester. Progress was easier on the plateau of the Karoo, and Pauling's men could advance as much as half a mile (1km) a day. Pauling pushed ahead despite opposition from the region's Afrikaner population, who hated the railroad as a symbol of British imperialism and as "an invention of the devil." But Pauling had an unexpected trump card to play. His family had welcomed in a couple of stranded Afrikaners who had been turned away by English hoteliers, thus ensuring the permanent gratitude and support of President Kruger of the neighboring South African Republic.

By 1890, Rhodes, a politician as well as an entrepreneur, had become Prime Minister of Cape Colony. A year earlier he had founded the British South Africa Company, which was to control the two countries later named after him—Northern and Southern Rhodesia. He planned to lay rails north to the Zambezi River and to the Nile Valley. The first step was Mafeking, 100 miles (160km) to the north of the existing end of the rails at Vryburg; the line was opened to traffic in October 1894. The next step was the 530 miles (850km) to Bulawayo. Pauling made good his promise to build the railroad at an amazing speed—more than a mile a day—and the line arrived at Bulawayo in 1897 to a banner reading "Our two roads to progress: Railroads and Cecil Rhodes."

The next step was to connect the Cape line from Bulawayo to Salisbury (now Harare in Zimbabwe), the capital of Southern Rhodesia. Two links were required: one connecting Salisbury to Beira on the coast of Portuguese-controlled Mozambique, and another running directly from Salisbury to Bulawayo. Construction was delayed by tension between the British and Portuguese authorities, which culminated in a diplomatic incident involving British forces and a Portuguese gunboat. The discord was resolved by a treaty between the two sides, and construction was permitted to proceed in 1892. Even by 19th-century standards it was a very risky project, crossing both swamp and forest terrain. In the first two years of construction more than half the white men died of fever, as did virtually all of the 500 Indian immigrant workers, who had less immunity to the local diseases. However, this did not deter Pauling & Co's project manager, Alfred Lawley, himself an excellent engineer, from achieving what Tabor describes as an "amazingly successful 2-feet [60cm] gauge miniature line, almost 'thrown together'... on the rough and ready earthworks," even though "at times it ran like a fairground switchback." The line would be improved a year later in 1899, when it was widened to the "Cape Gauge." The first train reached the Rhodesian frontier in February 1898 carrying the slogan "Now we shan't be long to Cairo." By 1902, the Bulawayo stretch of line was connected to Beira on the Indian Ocean, creating a continuous link of more than 2,000 miles (3,200km) to Cape Town on the Atlantic.

Meanwhile, there was considerable progress on the northern section of the railroad, which stretched south from the Mediterranean coast of Egypt. There had been railroad lines in Egypt since the mid-1850s, but financial and political problems ensured that they did not stretch into Sudan, the country's southern neighbor. In 1898, a major breakthrough took place when Major General Sir Herbert Kitchener, commander-in-chief of the British-controlled Egyptian army, arrived with an army to retake Khartoum, the capital of Sudan. The city had been captured 15 years earlier by the Mahdists, a rebellious local militia who had murdered General Charles Gordon and all the British inhabitants. To reach Khartoum from Egypt, Kitchener needed a railroad to convey the troops south from Wadi Halfa on the Nile, in order to bypass hundreds of miles of unnavigable sections of the river. Experts dismissed the idea as impossible, but Kitchener found his equivalent of Pauling in a much more orthodox character—a brilliant and experienced young French-Canadian railroad engineer called Percy Girouard.

Girouard identified a route that included a 250-mile (400-km) shortcut across the desert to Abu Hamed instead of following the winding river Nile, which took nearly 600 miles (1,000km) to reach the same point. It was not easy terrain, as the young Winston Churchill—who combined the roles of journalist and officer in Kitchener's army—explained: "it is scarcely within the power of words to describe the savage desolation of the regions into which the lines and its constructors plunged." Kitchener took the long view and decided that the line should be built using the "Cape Gauge," in view of the possible link up with Rhodes' line. Girouard established a veritable "railway town" at Wadi Halfa on the Nile, as well as a railhead—a mobile town complete with a station, stores, and a canteen—and reached half-way to Khartoum in a mere six months, on the same day that Pauling reached Bulawayo.

The line reached Atbara near Khartoum nine months later, in time to enable Kitchener to avenge Gordon at the Battle of Omdurman in September 1898. The battle was a virtual massacre that cost only 50 British lives, while several thousand Mahdi followers perished. As Churchill pointed out, such a war "was primarily a matter of transport. The Khalifa [the Mahdi's official title] was conquered on the railway." The conquest had a major political repercussion for the British Empire. Joseph Chamberlain, the imperialist Colonial Secretary, later told a reporter: "you will live to see the time when a railroad will be built through that country to the Great Lakes, the Transvaal, and the Cape."

BATTLE OF OMDURMAN, 1898
General Kitchener took bloody revenge for the killing of General Gordon by the Mahdist militia at Omdurman. His army reached the battlefield by train, on a purpose-built line.

"Build the bridge across the Zambezi where the trains, as they pass, will catch the spray of the Falls"

CECIL RHODES

Back in the south of the African continent, just before the Boer War broke out in 1899, Pauling had promised to fulfil Rhodes' dream to build a yet-more-ambitious line. He committed to extending the tracks from Salisbury to the Zambezi River at Victoria Falls, and then on to the Congo border at Likasi—more than 1,000 miles (1,600km) of track. It was projected that this track would take 14 years to lay. However, plans were delayed for three years by the war.

The railroads, commanded by the ubiquitous Girouard, proved to be vital for British communications during the conflict, but required a high proportion of British troops in order to guard the lines against attack. The first stretch of the new line was relatively simple, and the 300 miles (480km) of track across open savannah countryside arrived at the Zambezi in 1904. Soon the Zambezi Express from Cape Town was providing a regular service to the north. The seemingly impossible task of crossing the 650-ft (200-m) span of the Victoria Falls was achieved when the Cleveland Bridge & Engineering Company of Darlington in northeast England built a bridge to specifications set out by the British engineer George Hobson. It took five months to build, and was then shipped to the heart of Africa, where it was constructed on site.

The next extension of the line had a sound economic object: the enormous reserves of coal at Wankie, and the equally staggering riches of copper in the so-called Copper Belt at Broken Hill. Both areas were in Northern Rhodesia (now Zambia), and the route to reach them required a bridge even longer than that at Victoria Falls. Designed by Hobson, the structure that crossed the Kafue river had 13 steel spans and was completed in 1906 in a mere five months. Pauling and his colleagues had become even more adept at managing these huge construction projects, and completed the 281 miles (450km) from Kalomo, the existing railhead 50 miles (80km) north of the Zambezi, to Broken Hill (now Kabwe) in just 277 working days. The impetus behind the project, however, had been greatly reduced following the death of Rhodes in 1902 at the age of just 48. His

ZAMBEZI BRIDGE
The suspension bridge over the Zambezi River at the
Victoria Falls was an astonishing feat of engineering,
originally constructed in England and reassembled in
situ. Rhodes did not live to see it erected in 1905.

successor was Robert Williams, a Scottish mining engineer who
lacked Rhodes' British imperial vision. However, after three years
of negotiations, Williams obtained concessions from the Belgian
authorities to continue the railroad into the recently annexed
Belgian Congo. The line finally left British-controlled territory in
1909 on its way north to the Katanga region—which had even bigger
deposits of copper and other minerals than Northern Rhodesia—
rather than east to Tanganyika, which was a colony of German East
Africa at the time. It eventually reached Bukama, 450 miles (725km)
further along the Congo River, in 1918, the year before a defeated
Germany was divested of all of its East African territories.

By the end of World War I, the idea of a grand imperial railroad all
the way down the spine of Africa was abandoned. Instead, efforts
were concentrated on building the shortest route to export the
Congo's minerals to Europe; to this end, the Benguela railroad in
Angola was extended to Lobito Bay on the Atlantic. However, this
appallingly difficult route of more than 800 miles (1,200km) would
not be complete until 1929. Until then, a tenuous—and somewhat
roundabout—rail link did indeed run from the Cape to Cairo, also
using ferries to cross lakes and the River Nile. A few hardy travelers
succeeded in traveling along the entire route—that, surely, could
count as a success for Rhodes' vision, given the monumental scale of
the task of crossing the continent.

Cape to Cairo

A Victorian imperial ideal with the goal of opening the African continent up to commerce, Cecil Rhodes' ambitious plan for a north–south railroad across British colonial Africa was never fully realized. The challenges of the terrain, local opposition, and a lack of finance to meet the huge material demands of laying iron rails across mountain, jungle, and desert meant that, by the close of World War I, pragmatism had overcome idealism. Africa's abundant natural resources of diamonds, gold, and copper became the main destination of its railroads. This map shows the sections of the Cape to Cairo line that were completed between the 1880s and the 1920s, and its connecting railways.

ALGERIA

FRENC
WES
AFRIC

NIGERIA

KEY

- ● City
- ○ Town
- —— Main line
- —— Regional line
- ------ River route
- —— Colonial boundary

N
W ← → E
S

ATLANTIC
OCEAN

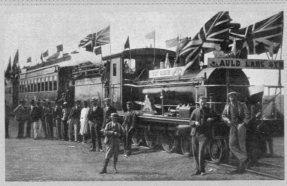

LINE OPENING
A branch of the Cape to Cairo line from Beira on the Mozambique coast to Salisbury (now Harare) in Northern Rhodesia (now Zimbabwe) opened on June 19, 1899. The arrival of the first train in Salisbury was celebrated by the line's engineers, George Pauling and Alfred Lawley, with British imperial pomp.

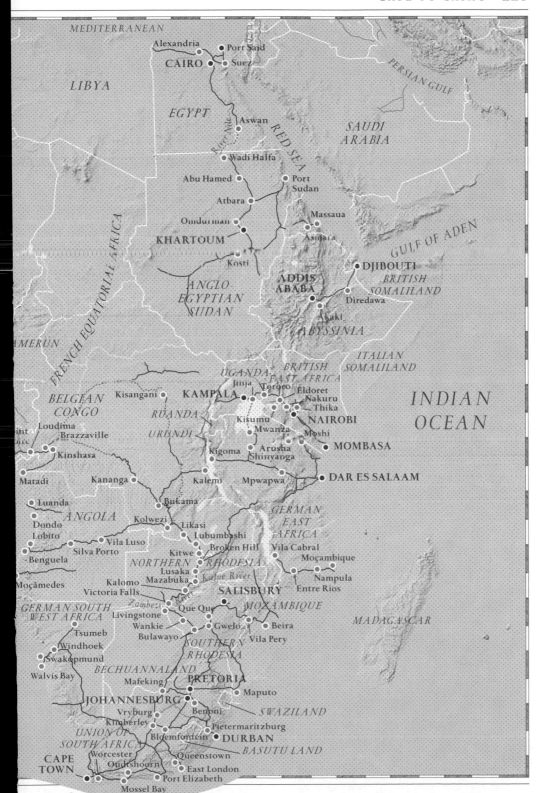

MEDITERRANEAN

LIBYA

EGYPT

Alexandria
Port Said
CAIRO
Suez

PERSIAN GULF

Aswan

River Nile

RED SEA

SAUDI
ARABIA

Wadi Halfa

Abu Hamed
Port
Sudan

Atbara

Massaua

Omdurman
Asmara

GULF OF ADEN

KHARTOUM

Kosti

DJIBOUTI
BRITISH
SOMALILAND

ANGLO
EGYPTIAN
SUDAN

ADDIS
ABABA

Diredawa

FRENCH EQUATORIAL AFRICA

Akaki

ABYSSINIA

AMERUN

ITALIAN
SOMALILAND

UGANDA
BRITISH
EAST AFRICA

BELGIAN
CONGO

Kisangani
Jinja
Tororo

KAMPALA
Eldoret
Nakuru
Thika

INDIAN
OCEAN

RUANDA

Kisumu
Mwanza
Moshi
NAIROBI

Loudima
Brazzaville

URUNDI

Kigoma
Arusha
Shinyanga

MOMBASA

Kinshasa

Matadi
Kananga
Kalemi
Mpwapwa
DAR ES SALAAM

Bukama

Luanda

ANGOLA
Kolwezi
Likasi

GERMAN
EAST
AFRICA

Dondo
Lobito

Lubumbashi
Vila Cabral

Vila Luso
Moçambique

Benguela
Silva Porto
Kitwe
Broken Hill

NORTHERN   RHODESIA

Nampula

Moçâmedes
Kalomo
Lusaka
Mazabuka
Kafue River
Entre Rios

Victoria Falls
SALISBURY

GERMAN SOUTH
WEST AFRICA
Zambezi
Que Que
MOZAMBIQUE

MADAGASCAR

Livingstone

Wankie
Gwelo
Beira

Tsumeb
Bulawayo
Vila Pery

Windhoek
SOUTHERN
RHODESIA

Swakopmund
BECHUANNALAND

Walvis Bay
Mafeking
PRETORIA

Maputo

JOHANNESBURG
SWAZILAND

Vryburg
Benoni

Kimberley
Pietermaritzburg

UNION OF
SOUTH AFRICA
Bloemfontein
DURBAN

BASUTU LAND

CAPE
TOWN
Worcester
Queenstown

Oudtshoorn
East London

Mossel Bay
Port Elizabeth

CAPE OF GOOD HOPE

Electricity
Lightens the Load

STEAM LOCOMOTIVES WERE ALMOST UNIVERSAL on the railroads by the end of the 1830s, once horses had been taken off the tracks. It was not until the end of the 19th century that electric power began to challenge steam's supremacy on the rails. Although electricity has numerous advantages over steam—it is cleaner, more efficient, and ultimately cheaper—it requires greater capital investment, as power has to be provided either by an external delivery system, most commonly an overhead wire or third rail, or by an internal system such as an onboard power generator—technologies that demand a high initial outlay. However, once its benefits had been seen, it superseded steam.

Britain was again a pioneer with electric trains, as it had been with the steam engine (see pp.22–29), although it lost out in the technological race once electricity was more widely adopted. As early as 1837, a chemist from Aberdeen, Robert Davidson, made the first electric locomotive. It was battery-powered, as was a second version named *Galvani* (after its galvanic cells, or batteries) that was exhibited at the Royal Scottish Society of Arts Exhibition in 1841. This massive, 7-ton (7.7-tonne) vehicle managed to haul a load of 6 tons (6.7 tonnes) at 4mph (6.5kph) for a

distance of 1½ miles (2.5km), and was tested on the Edinburgh and Glasgow Railway the following year. However, the perennial problem of batteries running out of power—which still limits their use in transportation today—meant *Galvani* was not much practical use. Railroad workers also opposed it, fearing electric trains would ruin their livelihoods, and destroyed the engine in a fit of Luddite pique.

German industrialist Werner von Siemens developed the first electric passenger train in 1879, and exhibited it on a 985-ft (300-m) circular track in Berlin, Germany. The train operated for several months, using a third-rail system to reach a speed of 8mph (13kph). Britain's first electric passenger railroad, the narrow-gauge Volk's electric railroad (named after its inventor, Magnus Volk), was completed in 1883 and, remarkably, still survives today, running 1¼ miles (2km) along Brighton's seafront. A low-voltage electric generator originally supplied a 50-V current to the small engine via the two running rails. Later the voltage was increased and the gauge widened from a narrow 2ft (60cm) to 2ft 8½in (80cm).

Some fascinating early experiments with electricity took place elsewhere. In Ireland, William Traill used hydro-electricity from a local waterfall to power a railroad for visitors to the Giant's Causeway tourist

SIEMENS' FIRST ELECTRIC PASSENGER TRAIN
Werner von Siemens' train was exhibited at the Berlin Trade Fair in 1879, to the excitement of visitors. Over a four-month period, 86,000 passengers took a trip on it.

BRITAIN'S FIRST ELECTRIC PASSENGER RAILROAD
High seas crash around the Volk's electric railroad along
Brighton's seafront. Opened in 1883, it carried 30,000
passengers in its first six months, and still runs today.

attraction. Traill built the 9¼-mile (15-km) Giant's Causeway, Portrush,
and Bush Valley Railway on a 3-ft (90-cm) gauge, and installed turbines
and dynamos to provide the power. However, electricity generation was
unreliable when the railroad opened in 1887, and steam engines were
used to supplement the electric power. Reliability improved once the
supply was converted to overhead wires rather than the third rail, which
was also hazardous to people crossing the line—in 1895, a cyclist died
from electric shock after touching the live rail. Despite its mixed success,
Traill's railroad was ahead of its day in ecological terms, and many lines
since, especially in the mountains, have been powered by more
sophisticated forms of hydro-energy.

The exception to the near universal adoption of steam for railroads
at their inception had been the trolleys—omnibuses that used tracks
through towns, so that they would not get stuck on muddy roads.
Steam engines trundling through towns would have been not only
dangerous but also impractical, as they operated poorly at slow speeds
when they had to stop and start a lot, so horses remained the sole form
of traction. Consequently, horse-drawn trams were the norm until the
electric tramways began to emerge—skipping a technological step.
The first commercial electric tram line opened in Lichterfelde, a suburb
of Berlin, in 1881. It was built by Werner von Siemens, who had exhibited

the first electric train two years earlier. In 1883, the Mödling and Hinterbrühl Tram—the first regular service in the world powered from an overhead line—opened near Vienna, in Austria.

In the US, electric trolleys (the American name for tramways) were pioneered in 1888, on the Richmond Union Passenger Railway in Virginia. The new technology encouraged rapid expansion and in just over 10 years, trolleycars had become almost universal across the country: there had been just 3,000 miles (4,800km) of horse tramways prior to electrification; by 1905, there were more than 20,000 miles (32,000km) of electric trolley lines and the trolleycar became the most common form of urban travel. Although there was a potential hazard associated with trolleys powered from overhead lines, which occasionally resulted in electric shocks, in practice this seems to have been rare.

Elsewhere, the increasing use of tunnels, especially in urban areas and through mountainous regions, stimulated the need for electric locomotives. Despite the success of London's Metropolitan Railway (see pp.130–37), which opened in 1863, it soon became apparent that steam engines in tunnels caused dangerous levels of smoke, leading local authorities to prohibit their use within city limits. Once again Britain, which had developed the first underground railroad, blazed the trail, with a line using electricity. The 3¼-mile (5.1-km) City and South London Railway, the world's first deep subterranean line, which opened in 1890, was bored through the London clay, so the whole railroad was below ground. Steam could not be used in the poorly ventilated tunnels. At first, cables were suggested as an alternative form of traction, but in the end electricity was used. Small engines provided the power, and at times they could not cope with the heavily loaded trains on the line, which had proved to be an instant success. As a result, it was not unknown for trains to fail to make it up the incline at King William Street, the terminus for the line in the City, and to have to roll back for a second attempt. Nevertheless, electricity rapidly became the power supply of choice for underground railroads and soon the power units were fitted under the passenger cars, eliminating the need for a separate locomotive. The early lines of the London Underground, which had been steam-powered (see pp.130–37), were all converted to electricity in the first decade of the 20th century.

The mountains and tunnels of Switzerland made it an obvious site for electric-powered trains. In 1896, the first commercial electric trains ran on the Lugano Tramway and, by 1899, a 25-mile (40-km) stretch of

AMERICAN TROLLEY
An electric trolley, or
streetcar, in Washington
DC, in 1895. Within
10 years of their
introduction, electric
trolleycars had become
the most popular form
of urban transportation
in the US.

main line between Burgdorf and Thun had been electrified. It was on
its mountain routes, however, that Switzerland pioneered electrification.
The Simplon Tunnel line was powered by electricity when it opened
in 1906 and the St. Gotthard line demonstrated the advantage of
electric traction when it was introduced in 1920. Whereas two steam
engines struggled to climb up the steep grades pulling a 200-ton
(224-tonne) load at 20mph (32kph), one electric locomotive could
haul a load of 300 tons (336 tonnes) up at 30mph (48kph). After that,
electrification became the norm in Switzerland and began to spread
rapidly across Europe. The Swiss success with electricity, combined
with a coal shortage after World War I and an abundance of cheap
hydro-electricity, drove its progress. Italy, which had already
electrified a couple of its mountain lines, and France both drew up
ambitious plans to electrify many of their main lines. Technical
problems combined with the resistance of railroad managers who
still favored steam held up France's program. Italy, however, rapidly
expanded its electrified services, driven by Mussolini, the nation's
dictator after 1925, who saw electric trains as epitomizing modernity.

The United States had just beaten Switzerland to become the first
country to electrify a main line when, in 1895, it opened a 4-mile
(6.5-km) stretch of the Baltimore Belt Line of the Baltimore and Ohio
Railroad—a connection from the main line to New York through a
series of tunnels around the edges of Baltimore's downtown. A 1903
decision by the New York State legislature to outlaw the use of smoke-
generating locomotives on Manhattan and in rail tunnels under the
Hudson River boosted electrification in the US. As a result, electric
locomotives began operation on the New York Central Railroad in

1906. In the 1930s, the Pennsylvania Railroad, which had introduced electric locomotives because of the regulation, electrified all its lines east of Harrisburg, Pennsylvania.

Despite the obvious advantages of electrification, the conversion of railroads to electric power remained patchy. This was partly because railroad managers were resistant to change, but also because it was hard for them to know which of the plethora of incompatible systems to choose. Many different kinds of electric technology were used: as well as various delivery systems, there was a wide variety of voltages, different phases (one , two , or three phase), and types of current   DC (Direct) and AC (Alternating) Even today, there are numerous systems in operation, which hampers the integration of services, especially across national borders.

The most difficult choice that promoters of electrification had to make was whether to use a third-rail or an overhead system. For the most part, overhead was used for mainline railroads, while commuter lines and suburban systems were generally equipped with a third rail. Sir Herbert Walker, the general manager of Britain's Southern Railway from 1923 to 1937, was a great pioneer of electrification in the country. He decided on the use of a third-rail rather than an overhead system—indeed, on the Brighton line, he replaced the existing overhead with a third-rail system. Nowadays, this is seen as outmoded and inefficient, as it relies on low voltages. However, with a network of more than 1,000 miles (1,600km) of suburban and regional railroad equipped with a third rail, it would now be prohibitively expensive to convert to overhead operation. The London Underground is unusual in that it operates on a four-rail system—one of the only ones in the world (the fourth rail helps to increase the total voltage available).

It was not until after World War II that new railroads were almost invariably powered by electricity and the majority of existing main lines in Europe were electrified. Oddly, however, despite its early adoption of electric-powered trains, very little of the US rail network is electrified today—a few passenger lines in the northeast, and some commuter services—while most mainline trains are powered by diesel engines. In India, too, many trains are still diesel-powered, but in most modernizing Asian countries, electrification of busy lines is now standard, and all high-speed trains around the world are electrically powered.

Going Electric

In the 19th century, electric trains were seen as a cleaner alternative
to smoke- and soot-belching steam locomotives. The first electric
trams appeared in the 1880s, with electric trains following in the
20th century. Cheaper diesel-electric engines arrived in the 1930s.

NER NO.1 (1904)
One of two Class ES1 electric locomotives
run by North Eastern Railway (Britain),
No.1 featured a "steeplecab"—a centrally
mounted cab—and could be powered by
overhead lines or an electrified third rail.

ENGLISH ELECTRIC NO.788 (1930)
A small locomotive used for switching at
the English Electric engineering works,
No.788 drew its power from rechargeable
batteries. This class was used for a range
of light industrial applications in Britain.

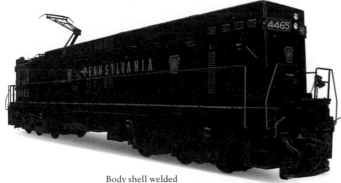

**PENNSYLVANIA
RAILROAD NO.4465 (1963)**
No.4465 was an E44-class
electric freight locomotive
built by General Electric.
A versatile and reliable
engine, the E44's six
traction motors could
produce a top speed of
70mph (112kph).

Body shell welded
rather than bolted for
streamlined look

Steeplecab design
for bi-directional
travel

**PENNSYLVANIA
RAILROAD
NO.4935 (1943)**
Nicknamed "Blackjack,"
No.4935 was a GG1-
class electric passenger
locomotive that was
later pressed into freight
service. Industrial
designer Raymond
Lowey's distinctive
streamlining hides a
powerful locomotive
with concrete ballast
for greater traction.

"Cat's whiskers" paint scheme

Three-axle truck

SNCF BB9004 (1954)
The BB9004 was a high-speed electric locomotive built for French national railroad SNCF. In 1955 it set a world speed record of 205mph (331kph) on the same day that another SNCF engine, the CC 7107, achieved the same feat. The record was not beaten until 2006.

DB CLASS 160 "BUGELEISEN" (1927–34)
German national railroad Deutsche Bahn operated a fleet of 14 Class 160 electric switching locomotives between 1927 and 1983. Its distinctive appearance led to the nickname *bugeleisen* ("iron"), due to the placement of the cab behind the engine.

CONRAIL NO.2233 (1963)
Consolidated Railroads arose in 1976 out of the wreckage of six failed railroad companies. This general purpose GP 30 diesel-electric locomotive was one of the "second generation" of diesel engines, and was first unveiled by the Electro Motive Division of General Motors.

JRF FREIGHT EH200 DC (2001)
The EH200 class was an electric freight train built by Toshiba for JR Freight, Japan's main cargo carrier. With a top speed of 70mph (110kph) and a power output of 6,061hp (4,520kw), its primary uses are for hauling oil tanks and working steep grades of track.

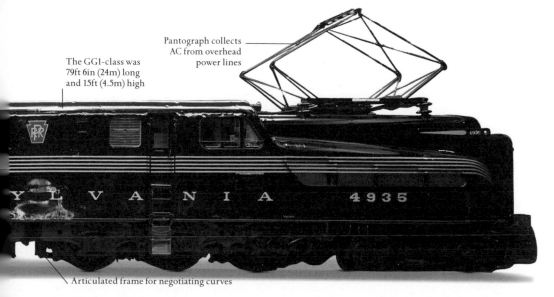

Pantograph collects AC from overhead power lines

The GG1-class was 79ft 6in (24m) long and 15ft (4.5m) high

Articulated frame for negotiating curves

The People Who Ran the Railroads

RAILROADS NEEDED MASSIVE NUMBERS of people to keep them running. Once the tracks were laid and the labor force of navvies had dispersed, a whole range of new personnel was required: engineers, firemen, porters, conductors, brakemen, maintainers, track workers, and tower operators. From their inception, many railroads were responsible for the largest workforce in their country. During the early days, and in many countries well into the 20th century, this workforce was almost exclusively male, and it was only during and after the World Wars that women began to be employed on the railroads in any number.

When the Liverpool and Manchester Railway opened in England in 1830 (see pp.25–29), it was not only the first commercial passenger railroad line in the world, but it also introduced many standards and practices that other railroad companies all around the globe would follow. Strict discipline and attention to detail were vital in order for the trains to run on time and to ensure the safety of the new railroads, so employees were expected to follow military-style rules and wear smart uniforms. Since the armed services were the only contemporary organizations comparable in size and scale to the railroads, inevitably their ex-personnel became a major source of trained, well-disciplined labor. Indeed, many of the first workers who took up posts as porters, engineers, firemen, and track workers were former soldiers or sailors.

As the railroad business boomed in the mid-19th century, running the increasingly large companies required new skills and new types of professional workers, such as specialist engineers, accountants, lawyers, and managers. Many of these workers were also recruited from the armed services, but this time from the officer ranks. This did nothing to lessen the military character of the early railroads, and issuing "orders of the day"—setting out the daily tasks to be undertaken— became a popular practice at this time. However, the railroads were not solely the domain of military men; white-collar jobs such as ticket-office clerks were snapped up by anyone with a modicum of education, and the rapidly expanding industry was so different from any other that training mostly consisted of learning on the job.

STATION MASTER, 1904
The role of station master was a much-sought-after white collar job. This station master at Finmere Station, Oxfordshire in England would have been an important, well-respected figure in his local community.

In many countries, the structure of the railroads simply reflected the dominant social or political hierarchy. When India was under British colonial rule (pre-1947), the laborers and unskilled staff were native people, but managers were predominantly white Europeans, or Eurasians with an Indian mother and a European father. Recruiting people to work on the railroads was more difficult in parts of the world where the tracks traversed inhospitable and sparsely populated territory. On the Trans-Siberian Railway, completed in 1916 (see pp.180–89), some of the local workforce consisted of convicts who had been sent east to Siberia as punishment, but the railroad company could not afford to be too selective in its recruitment. Thus, many of the security guards hired to guard property at night had been banished to Siberia for robbery, passengers were often served by conductors and ticket clerks who had committed violent crimes, and large stretches of track were maintained by murderers and rapists.

The rules of the railroads were strict and workers would be fired instantly for serious offenses, such as falling asleep on the job, or have their wages docked for more minor transgressions, such as "deserting" their post to get a cup of tea. It was hardly surprising, however, that some people fell asleep at work: workers toiled for up to 16 hours a day, six days a week. Furthermore, some of the restrictions placed on railroad workers were notoriously harsh. Engineers on delayed trains received no extra pay, or time off in lieu, even if they worked several extra hours. Employees could be fired without any warning, but they were required to give employers three month's notice if they wanted to

PENN STATION LUGGAGE, 1910
Porters were a vital part of the early railroads. They carried out of a variety of duties, including carrying the passengers' luggage.

leave. A London and South Western Railway Company employee who failed to give the required notice was prosecuted and sentenced to three weeks' hard labor in a case that served as a warning to others.

Discipline was largely the domain of special police forces, employed by individual railroad companies. A railroad policeman's duties included keeping order at stations, on the railroad, and in the local area around the railroad; removing trespassers and protecting railroad property. Most policemen carried a billy club and gun as weapons.

In most countries, working for the railroads was highly regarded, especially at the beginning of the railroad age. Even unskilled men could earn relatively high wages—railroad laborers earned twice as much as farm laborers. The relatively good pay, strict rules, and the stylish uniform gave the new industry prestige and afforded its workers a certain level of respect. Corporate loyalty was also strong as working for the railroads offered long-term stability, a permanent job in a world where that was a rarity, particularly in rural areas. Many men spent their whole working lives in the railroad industry and jobs were often kept within families for several generations. Railroad companies were usually keen to employ multiple members of the same family because they saw it as another way of fostering loyalty. The English railroad companies also devised other ways of keeping their employees happy: many jobs required working in remote areas so housing was provided at very reasonable rents. In towns, too, where workers had to start very early, housing was often provided near depots and stations. Providing cheap, convenient housing meant that workers would be reluctant to leave their jobs since it would make them, and their families, homeless. Moreover, the most efficient and most loyal workers were rewarded with the best homes. As Frank McKenna puts it in his history of

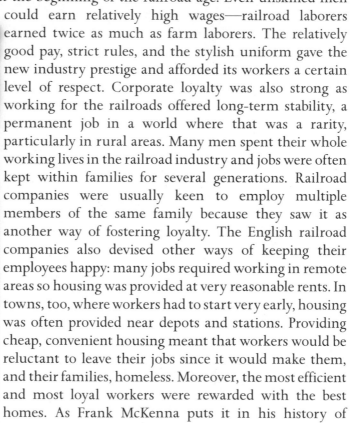

POLICING THE RAILS
Railroad police certainly looked the part; they wore stylish uniforms and had military-style ranks. Police officers carried billy clubs featuring the crest of the railway company, such as the Midland Counties (England) left. Senior police, such as detectives, carried a brass or ivory staff with a crown at the tip.

railroad workers, "from the earliest days, the companies used housing policy as a means of staff control and for the preservation of company loyalty." Free or concessionary fares for employees and their families were also a widespread benefit in kind that still exists today.

The railroad companies, therefore, took a paternalistic approach, controlling their employees with a firm but seemingly generous hand. However, railroad work was often very dangerous (only mining and fishing had a higher casualty rate), although most employers did not feel inclined to address the problem. The most hazardous jobs were switching and coupling or decoupling cars, as this involved working on the track right next to moving trains, but all track workers faced a high risk of being hit by a train. Moreover, in the early days of the railroads, individual cars didn't have brakes, so when the engineer stopped the locomotive, the cars simply bumped into each other to stop. On average, more than five times as many workers were killed in railroad accidents as passengers. In Britain in the first half of the 1870s, for example, there was an average of 782 worker deaths per year. In the US, the number of fatalities was even greater, with more than 2,000 dying in 1888 alone, a toll which belatedly led to the sponsoring of the Railroad Safety Appliance Act in 1893, which gradually began to reduce the number of accidents. Eventually, the loyalty of the railroads' employees was simply pushed too far. Concerns over safety, wages that hadn't kept pace with other industries, and autocratic management led railworkers to join

"The great object… is, to place the two rival powers of capital and labor on an equality so that the fight between them, so far as fight is necessary, should be at least a fair one"

BRITISH PRIME MINISTER HENRY CAMPBELL-BANNERMAN
ON THE 1906 TRADES DISPUTES ACT

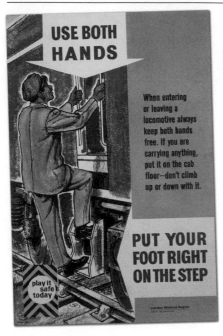

USE BOTH HANDS

When entering or leaving a locomotive always keep both hands free. If you are carrying anything, put it on the cab floor—don't climb up or down with it.

PUT YOUR FOOT RIGHT ON THE STEP

play it safe today

London Midland Region

CLIMBING THE LADDER
During the latter part of the 20th century, health and safety at work became a serious issue. As trade unions grew, the welfare of workers became a major concern at last.

together to form trade unions, which could coordinate campaigns for improvements to pay and conditions. The traditional loyalty that workers felt toward the railroad companies began to erode. In Britain, trade unions started to be organized in the 1860s. In 1867, workers on the North Eastern Railway went on strike over working hours—they wanted the company to agree to a maximum ten-hour-day or sixty-hour-week, which does not seem unreasonable by modern standards. However, the company reacted aggressively by employing strikebreakers and firing the strikers, so the strike collapsed.

However, trade unionism could not be held back for long and gradually organized labor movements gained a foothold over the ensuing couple of decades. Membership of unions grew quickly, and employers had to take notice. The unions were able to wrest a few concessions from the railroad companies, principally over the long hours and lack of overtime pay, but the companies were far from happy about it. In Britain, matters came to a head on the small Taff Valley Railway, a line serving several coal mines in South Wales. The Amalgamated Society of Railway Servants union had successfully negotiated a 60-hour-week in 1890, following a brief strike, but the Taff Vale Railway Company still failed to recognize the union officially. This provoked a second strike by the workers in 1900, but the company decided to sue the union. In 1901, the strike was judged to have been illegal and the union was ordered to pay £42,000 ($205,000) in compensation to the company, a judgment that made any further industrial action in Britain impossible. This incident provoked a bitter public reaction and as a result the Conservative government was heavily defeated in the 1906 general election. The new government, led by the Liberals, introduced the

Trades Disputes Act, which effectively gave trade unions immunity from being sued in such circumstances. This resulted in a rapid growth of union strength.

Other countries followed a similar pattern. In the US, railroad workers began to organize into trade unions in the 1860s, but this was strongly opposed by the railroad companies. Railroad workers were involved in three major strikes in the last quarter of the 19th century and although all of them ended in defeats for the unions, union membership nevertheless increased and the rail companies were forced to recognize them. The US public largely supported the workers' desire for basic rights and in the early years of the 20th century trade unions grew across the rail industry, and they remain relatively strong in the 21st century. Similarly, in the Netherlands, railroad companies were initially resistant to any legislation that would restrict the number of hours that employees worked. So, in the late 1890s, brotherhoods began to form and in 1901 they coalesced into a single trade union, the Federatie van Spoorwegorganisaties. The union's first strike, in January 1903, was in support of dockworkers and immediately won concessions, but a second strike in April over working conditions resulted in mass firings. Eventually, though, hours were reduced and workers' wages increased.

Today, railroads require far less labor. Interlocking towers have been replaced by rail traffic control centers covering vast regions; electric and diesel trains only require a single engineer; stations rarely employ porters; ticket sales are often automated; and machines have replaced people for some track safety and maintenance tasks. Some modern subway systems are even operated without engineers, with trains controlled by computers. The days when the railroads were the largest and most prestigious employers are largely over, although Indian Railways is still the ninth largest employer in the world. However, with more than 620,000 miles (1,000,000km) of railroads around the world, a significant, skilled workforce remains vital to build, maintain, and operate the railroads.

The Wrong Side of
the Tracks

FROM THE 1830s ONWARD, railroads became potential sources of huge profit. People and goods could be transported further, faster, and for less money than ever before. Those local merchants who had invested in the earliest railroads, such as the Liverpool to Manchester line (see pp.25–29), reaped the financial rewards. Further railroad expansion was needed, and many people saw it as the perfect opportunity to make lots of money. By the 1840s, the rapidly expanding railroad system was attracting many new businessmen, entrepreneurs, and investors, all hoping to see fat returns for their money, if not always by honest means.

In 1859, a pioneering investigative journalist, D. Morier Evans, described the situation in his book *Facts, Failures and Frauds*: "It is with the railway mania of 1845 that the modern form of speculation may be said to begin and the world has not yet recovered from the excitement caused by the spectacle of sudden fortunes made without trouble." It was finance capitalism at its most primitive, with George Hudson (see pp.55–55) typical of those who, in Evans' words, were "pioneering new lines through every difficulty" and establishing the many and various ways in which he and his successors could defraud the public. Hudson embezzled a fortune and ruined many investors along the way, including the famous English literary family, the Brontës.

By modern standards of corporate governance, the promoters responsible for many of the world's railroads were dishonest. Yet many of them, such as Henry Meiggs in South and Central America

"A man who has never gone to school may steal from a freight car; but if he has a university education, he may steal the whole railroad"
THEODORE ROOSEVELT

(see pp.198–203), made innovative contributions to the development of the railroads. Even George Hudson belonged to the class of railroad promoters who actually cared passionately about railroads, as well as feathering their own nests. So perhaps the worst scoundrels were those who contributed nothing to the actual development of the railroads. John Sadleir, a successful Irish financier and Lord of the (British) Treasury, who issued £150,000 (around $750,000) worth of forged shares in the Royal Swedish Railway Company, certainly belonged to the latter category.

While John Sadleir eventually ruined himself, his railroad fraud ensured that the Swedish government assumed responsibility for its country's rail system for the next 150 years. In this the Swedes were typical of many European governments whose control over their railroads was far more complete—and more honest—than the British. The French, for instance, operated a system of controlled regional monopolies. Nevertheless there remained opportunities for those involved to enrich themselves, and some politicians often made handsome, if completely unethical, profits from the railroads. In Prussia (later part of Germany), Chancellor Otto von Bismarck had no qualms about instructing his banker to buy shares in the railroads he was proposing to nationalize—a process that made him a tidy profit, but that would now be deemed illegal insider trading.

Perhaps inevitably, it was the US, with the biggest rail network in the world (at that time), that provided crooks with the greatest opportunities. The battle for control of the Erie Railroad in the late 1860s really laid bare the world of railroad speculation—the varied characters, the complex legal dealings, and the level of political involvement. The battle was immortalized by Charles Francis Adams, a grandson and great-grandson of United States presidents and himself a railroad man, in his book *Chapters of Erie* (1871). The Erie Railroad had never been profitable, he explained, but industrialist Cornelius

JOHN SADLEIR
John Sadleir was immortalized by several writers, including Charles Dickens and Anthony Trollope, who are believed to have based on him the characters of Mr. Merdle in *Little Dorrit* and Melmotte in *The Way We Live Now* respectively.

"Commodore" Vanderbilt saw an opportunity to create a monopoly of the tracks to Lake Erie and make some money. However, he came up against a formidable trio of directors and speculators—Jay Gould, who had acquired a reputation as the most sinister and corrupt of railroad barons, and his two associates, Daniel Drew and Jim Fisk. The Gould gang issued masses of stock—and bonds convertible into stock—in order to dilute Vanderbilt's holdings. The ensuing battle spread to the courts and the politics of New York State. When the battle reached the State legislature in Albany, Vanderbilt looked like he was winning the hearts, minds, and pockets of the legislators, but Gould got there first—at an estimated cost of $1 million. In the end, Vanderbilt gave up.

ERIE SATIRE
Puck, a popular American magazine, published this cartoon satirizing the struggle for control of the Erie Railroad. Vanderbilt gleefully watches Jay Gould drowning in a flood of his own watered-down stock.

Russia, too, was ripe for corruption. During the railroad boom in the latter part of the 19th century, many lines were built by private enterprise with the government guaranteeing a generous rate of return on their investment. This proved highly lucrative. One of the big railroad barons of the Russian railroads, Samuel Polyakov, manipulated the companies he ran to ensure that he owned all the shares, thereby reaping all the dividends. Polyakov also amassed shares in other railroads, which he used as collateral against loans from foreign bankers, betting on the expected rise in share value. While these activities might just have been on the right side of the law, he also artificially inflated costs of railroad construction in return for bribes to state officials, usually paid with railroad shares. And he was not the only one. In *A History of Russian Railways* J.N. Westwood writes: "Many other important civil servants and even members of the royal family (including apparently Tsar Alexander III's brother, the Grand Duke Nikolai Nikolaevich) received bribes in the form of shares from railway promoters." This corruption was endemic at all levels—even conductors, who were poorly paid, would often allow a passenger to travel for a "consideration," usually around half the proper fare.

German philanthropist Baron Maurice von Hirsch died both respected and respectable, yet he was also a railroad profiteer. Today, his fortune, most of which he made on the Constantinople-to-Vienna railroad, would be counted in billions. Hirsch was awarded a concession by the Sultan of the Ottoman Empire to build lines in the open country, which did not touch any towns. This left gaps that had to be filled, which Hirsch did—expensively—and he was reckoned to have made several million dollars from the construction contracts alone. To make matters worse, the lines were built so shoddily that they required extensive improvement. Hirsch increased his profits by issuing "loans" to himself at cheap rates, then selling them on to banks at a profit. (The banks then sold them to the public even more expensively.)

Every aspect of railroad promotion, construction, management, and operation attracted its own type of crimes and criminals. Unsurprisingly, given the crowds, stations were favorite haunts of pickpockets and other opportunistic rogues. The managers of the London and North Western, Britain's largest railroad company, were, according to the historian of the line, "at their wits' end to find out the blackguards. Not a night passes without wine hampers, silk parcels, drapers' boxes or provisions being robbed; and if the articles are not valuable enough they leave them about the station." In the US, cardsharps proliferated on the railroads, including the infamous "Poker Alice" Ives, a petite, blue-eyed beauty whose winning combination of card skills and feminine wiles made her a very successful gambler.

Access to the files of a railroad company also provided rich pickings for forgers and con artists. None was more successful than Leopold Redpath, a British clerk at the Great Northern Railway, the company that owned the tracks between London and York. Redpath was assisted in his deception by the fact that the company had numerous kinds of stock, bearing various rates of dividend, and therefore intricate calculations were required to determine the level of payments to which the owners of the certificates were entitled. Redpath forged share certificates in which he was both buyer and seller, or, more cunningly, added the figure "1" to genuine documents bought by him, transforming, say, £250 to £1,250. In total, Redpath is believed to have embezzled an astonishing £220,000 (about $29 million in today's money) through forgery and speculation on the company's stocks and shares (which was also forbidden). Interestingly, although he used his ill-gotten gains to live a comfortable and luxurious life, Redpath also became a philanthropist; as Evans summarizes:

> never was money obtained with more wicked subtlety; never was it spent more charitably. A greater rogue, so far as robbery is concerned, it were difficult to find; nor a more amiable and polished benefactor to the poor and the friendless.

Ultimately, it took the revelation of anomalies in the affairs of another business, the Crystal Palace Company, to prompt a similar investigation at the Great Northern Railway. This lead to the discovery of Redpath's deception and resulted in his conviction and banishment to a penal colony in 1857. Redpath was not alone, however; his defense counsel revealed that when Redpath joined the Northern "he found in the office of the company a widespread system of speculation and of trading in stocks and shares

WILLIAM ROBSON STOLE

$132,000

(£27,000) FROM THE CRYSTAL PALACE COMPANY IN 1855

under other people's names—names not infrequently entirely utterly fictitious." It is impossible to know how many similar frauds occurred during this time, but have remained undetected.

One railroad crime from this era stands out for its cunning execution—the theft of £12,000 of gold bullion (around $1.6 million in today's money) in 1855 from a South-Eastern Railway train traveling from London to Folkestone, Kent. It was largely an inside job involving Messrs. Burgess, a guard, Tester, a clerk, and Pierce, a ticket-printer and con artist, together with Edward Agar, a long-time professional thief. The plan was devised by Pierce and Agar: the gold was bound for France, and would be transferred from the train to the ship at Folkestone. On the way to Folkestone, it was carried in a guard's van in safes with two separate locks and was then weighed at the docks to ensure that nothing had been stolen. The gang managed to acquire impressions of both sets of locks and enough lead to counter the weight of the gold. They were able to board the train, open the safes, extract the gold, and replace it with the lead, all before the train reached Folkestone. It was so wonderfully simple and the scheme only unraveled when Agar was arrested for an unrelated crime. He asked Pierce to provide for Fanny Kay, the mother of one of his children, but Pierce failed to do so. Kay then revealed the plan to the railroad company and Agar confessed, implicating the other three to reduce his own sentence.

Thus the early commercial railroads attracted a wide range of crooks and cheats, forgers and con artists, speculators and schemers from every echelon of society, all looking to line their pockets while emptying others'. As Samuel Smiles put it in his 1857 biography of "father of the railways," George Stephenson: "Folly and knavery were for a time in the ascendant. The sharpers [cardsharps] of society were let loose and jobbers and schemers became more and more plentiful. They threw out railway schemes as lures to catch the unwary."

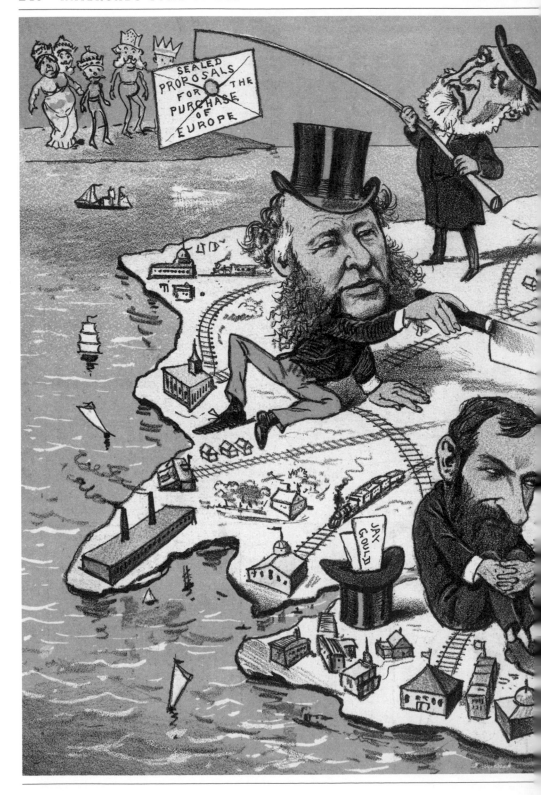

WHAT A CARVE UP!
Entrepreneurs Cornelius Vanderbilt, Jay Gould, Russell Sage, and Cyrus W. Field are shown dividing up the United States' railroads in 1882, as European royalty watch from across the Atlantic Ocean.

Indian Hill Railroads: Climbing Out of the Heat

THERE WERE MANY REASONS to build railroads: to carry passengers and freight; to unite towns, villages, and nations; to make money; or simply to stop other companies from building them. The famous Indian hill railroads, however, owe their existence to a very particular phenomenon—the British colonists' dislike of the hot Indian summer.

Since their inception in 1853, the railroads in India had quickly become a vital part of the way of life in the subcontinent, used by both the Indians and the British, though mostly in separate cars or trains. As the railroad network grew, the British realized that it could also be the answer to the nagging question of how to avoid the oppressive heat during the summer months. Leaving the towns in the summer for cooler areas in the hills had long been a habit of the colonists, but it was a long and arduous journey. Building the Indian hill railroads, which would climb perilously steep inclines to link the hill towns with the plains below, seemed the obvious, if ambitious, solution. Such a venture seemed impossible at first, however, as the hills in question—which included the foothills of the mighty Himalayas—had grades that appeared too steep to tackle. Yet as the 19th century wore on and railroad engineering became more and more sophisticated, it was soon felt that no mountain, ravine, or river could present any serious obstacle to the iron road's inexorable progress.

The first, and still the most famous, hill railroad to be built in India was the Darjeeling Himalayan Railway (DHR). It linked Siliguri in the plains of the Himalayas (400ft/122m above sea level) with Darjeeling in the Lesser Himalaya mountain range (6710ft/2,045m above sea level), climbing nearly 1¼ miles (2km) in its 55-mile (88-km) journey. Construction began in May 1879, just months after the mainline railroad had reached Siliguri, and it aroused great interest in India. In March 1880, Lord Lytton, the British Viceroy of India, traveled on the first completed stretch of track, cheered on by huge crowds. The line was completed in two years, which is truly remarkable considering the challenges the engineers faced. For most

CHUNBATTI RAILROAD LOOP, 1914
Originally the third loop of the Darjeeling Himalayan,
the Chunbatti loop is now the first and lowest after
the others were removed in 1942 and 1991.

of its length, the DHR ran alongside a newly built path, but while the first 7 miles (11km) was a gentle incline, after that the grade became much steeper, up to 1 in 23 (or 4 percent). This level of incline can be problematic for a railroad without extra support, such as a rack or a cable, but the DHR engineers avoided this by building a lighter, narrow-gauge railroad. Later they also added loops, in which the track passed over itself, and switchbacks, in which it reversed back on itself, to reduce the sharpness of the grade further. One of the four loops on the DHR was named Agony Point because of the perilous tightness of the bend and its proximity to the precipice of the hill.

Perhaps surprisingly, the DHR was extremely profitable right from the start. Not only did it carry British residents eager to escape the summer heat and tourists who immediately flocked to see the wonderful scenery, it also carried vast amounts of tea. In fact, the arrival of the railroads helped the local tea industry to thrive. Although the railroad was very slow, rarely reaching speeds of more than 15mph (24kph), it was still much faster than the bullock carts that used the road. As with the mainline Indian railroads, the hill railroads

AGONY POINT, 1910
The fourth loop on the Darjeeling Himalayan
Railway, Agony Point also has the tightest
curve. It is located just north of Tindharia
station, on the steepest part of the line.

also had a military function: building railroads deep into the Himalayas enabled the British to establish control and create garrison towns that would protect the most remote parts of the subcontinent.

In the early days, the mail train left Siliguri at 8:25am and passengers were treated to an unparalleled experience as the train climbed up through the cloud and early morning mist, steadily navigating the steep grade to find the warmth and blue skies of Darjeeling. An early description in *Railway Magazine* in 1897 recounts an amazing climb up the Himalayan foothills:

> And now we approach the culminating wonder of the line. At one place, we have been able to count three lines of rail below us which we have just traversed, and to see three more above us, up which we are going to climb, making in all seven lines of track (counting the one we are on) visible at one time, nearly parallel with each other at gradually rising heights on the mountain side. But now the wheels groan with the lateral pressure caused by a tremendous series of curves, and for a few breathless seconds, the train seems transformed into a veritable snake, as we pass 'Agony Point' and in so doing traverse two complete circles of such incredibly small diameter that the train, if at rest, would stretch round more than half of the circumference of one of them.

Today, the DHR is a designated UNESCO World Heritage Site, honored for its innovation and socioeconomic impact. It is also still a functioning railroad, carrying both local passengers—there is even a special school train, which transports children to schools in local towns—and countless tourists. The trains can be heard anywhere on the mountain up to Darjeeling, thanks to the loud horns that sound constantly, even managing to drown out the trucks and buses. If they are lucky, passengers can see Mount Everest in the distance, although the frequent mists and clouds make this

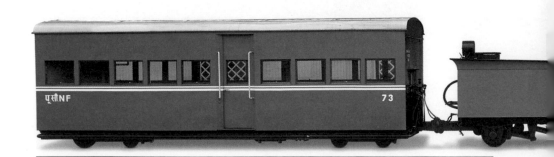

relatively rare (this author spent a week there without so much as a glimpse of Everest). Incredibly, some of the original steam locomotives supplied by the British company Sharp, Stewart & Co. are still running, although they spend much of their time in the workshop at Tindharia, a third of the way up the line. The line has suffered from subsidence and landslides over the years and the lower section was closed in 2010 following a major landslide, which also washed away the adjoining road. Today, only the upper section from Kurseong to Darjeeling remains in operation.

The next hill railroad project in India was the 3ft-3⅜-in (1-m) gauge Nilgiri Mountain Railway (NMR) from Mettupalayam to Udhagamandalam (more commonly known as Ooty) in Tamil Nadu, South India. It had initially been proposed as a garrison line way back in 1854 at the start of the Indian railroad age, but construction did not begin until 1894. The NMR was even steeper than the DHR, as the railroad had to climb more than 1 mile (1.6km) from the plain to the summit, over a distance of 26 miles (42km). Consequently, it took much longer to build than the Darjeeling line and entailed far more major structures, including 108 curves, 16 tunnels, and a staggering 250 bridges. It did not open fully until 1908. The steepness, which at some points reached 1 in 12 (or 8 percent), was too great for a locomotive to manage on its own and consequently for the first 17 miles (27km), between the towns of Mettupalayam and Coonoor, the line used a rack and pinion system to navigate the steep grade. This involved installing a third rail in the middle of the track, with teeth similar to those on a gear mechanism. A special toothed wheel on the locomotive then locked in to the rail, enabling it to grip and help to pull the train up the hill (see pp.108–109). Given the steepness of the line and its sharp curves, it was a very slow journey, taking five hours to reach Ooty at the top. As with the DHR, some very old steam locomotives still

DHR TRAIN
Due to the railroad's narrow gauge and the distinctive blue trains that resembled beloved fictional steam locomotive Thomas the Tank Engine, the Darjeeling Himalayan Railway also became known as the "Toy Train."

travel on the line today. In honor of that fact, the NMR joined the DHR as a UNESCO World Heritage Site in 2005, and they became known collectively as the Mountain Railways of India.

The third famous Indian hill railroad to be completed was the Kalka to Shimla line, opened in 1906. Shimla (or Simla as it was then known) was a much more important center for the British than Darjeeling or Ooty. By the 1830s, it had become a well-established summer residence for the British and was noted for its balls and other social highlights, attended by colonial officers and senior administrators. Although the roads up to the station were widened and improved with the construction of the Hindustan–Tibet highway in 1850, it was still a four-day journey

RACK AND PINION
The Nilgiri Mountain Railway adopted a Swiss-designed rack and pinion system, involving a third rail in the center (see below) to cope with the steepness of the grade.

from the plains up to Shimla. When, in 1863, Shimla became the official summer capital of India it meant that the whole paraphernalia of government, even the military, was moved between Calcutta and Shimla twice a year.

Due to Shimla's importance to the colonists, it was vital that a railroad should be built up to the town, but it was a far more

WHEN IT IS EVENTUALLY COMPLETED THE KASHMIR RAILWAY WILL EXTEND

214 MILES

(345 KM)

complicated proposition than the other two hill railroads. As with the DHR and the NMR, the Kalka to Shimla Railway was built to a very narrow gauge (in this case 2ft 6in/762mm) to make it lighter and to save construction time. Nevertheless, the railroad route from Kalka to Shimla involved the construction of more than 806 bridges and, although many were little more than culverts over streams, several were arched, multitiered structures rising from the bottom of deep valleys.

In fact, there is a sad legend about the longest tunnel on the line (number 33), which is nearly ¾ mile (1.2km) long. Colonel Barog, the engineer of this tunnel, ordered digging to begin at both ends, but the two sections were not properly aligned so they didn't join up. Barog was fined a nominal amount (1 rupee) by his employers, but he could not cope with the stigma of his failure and committed suicide. Another engineer completed the tunnel, but it was named the Barog Tunnel in memory of the original engineer. Despite this minor mishap, the 60-mile (97-km) line was a feat of engineering and came to be known as the "British Jewel of the Orient." The prolific railroad writer O.S. Nock traveled on the line in the 1970s and was amazed at the difficult terrain that the line traversed. He wrote:

> The geology of the area is highly erratic. The formation consists of a heterogeneous mass of boulders, clay containing small quantities of sand and other debris, while in other locations it is a solid rocky mass. There is frequent trouble during the monsoon season from slips and subsidence, and these sometimes occur without any preliminary warning because of the peculiar geology and the unpredictable hydrology of the area.

He goes on to explain that he found the use of the word "erratic" by local geologists rather strange until he traveled up the line and saw for himself that the area posed particularly unique difficulties for the construction of a railroad. It was worth it though. The line is one of the most impressive in the world with spectacular views of the Himalayan foothills and in 2008 the line was added to the UNESCO World Heritage Site, the Mountain Railways of India.

Several other hill railroads were built during this era and are still thriving today. These include the Kangra Valley Railway (opened to passengers in 1929) in the sub-Himalayan region; the Matheran Hill Railway (built in 1907), which ascends the western Ghats mountain range in south; and the Lumding to Silchar line, built at the turn of the 20th century, lying deep inside the state of Assam, in the Barak river valley of the Cachar Hills. All of them are spectacular railroads, providing a vital connection between remote hill towns and the lowlands. However, Indian hill railroad-building is not just a thing of the past: the Kashmir Railway, which aims to connect the region in the outer Himalayas with the rest of the country, is an ongoing challenge. First suggested in 1898, the proposed route includes major earthquake zones, extreme climates, inhospitable terrain, and is subject to continued conflict between India and Pakistan over territorial rights to Kashmir. These geographical and political difficulties combined to delay the start of building work until the late 20th century and completion is not estimated to happen until 2017, well over a century after the line was first proposed.

KALKA TO SHIMLA RAILWAY
As well as more than 800 bridges of varying sizes and complexity, the Kalka to Shimla line also featured 107 tunnels, a figure which has since been reduced to 102.

War and Uncertainty

MALLARD, NO.4468
LNER Class A4 PACIFIC
STEAM, 1938

The railroads reached their apogee just before the outbreak of World War I. Trains were now safer, cheaper, and faster than they had ever been, and reached virtually every sizeable town and village in the developed world. The United States was the world's leader, with more than 250,000 miles (400,000km) of line. Inevitably, it was the railroads that bore the brunt of coping with the huge transportation demands of the war. Not only did they take virtually all the war materials to ports for dispatch overseas and to the front, but they also carried millions of men off to war, and brought the casualties home. Indeed, the armies built whole networks of narrow-gauge lines to carry men and supplies right up to the front line. Railroads inevitably became targets during the conflict, notably in the Middle East, where Lawrence of Arabia led a series of assaults on the Hejaz Railway controlled by the Ottoman Empire.

After the war, the railroad companies began to realize that they had to look for alternatives to steam locomotion. There had been some electrification before the war, but now diesel was being considered, and both the Germans and the Americans created fast new diesel services in an effort to improve journey times. Attempts were still made to modernize steam locomotives, however, and the fastest ever speed by a steam engine was reached in 1938 by the British locomotive *Mallard*.

The railroads remained prominent in World War II, in which they saw their darkest hour—millions of Jews and other minority groups were transported by train to the German concentration camps, and thousands of prisoners died while constructing the Burma–Siam railroad. Elsewhere, they played a vital part in the eventual Allied victory over the Germans and the Japanese. However, the railroads were poorly treated after the war—as soon as hostilities ended, services closed in many countries as competition from road transportation and aviation intensified.

The Golden Age of the Railroads

IN MANY WAYS, THE IDENTIFICATION OF a "golden age" of rail is difficult. At various times there were many incredible railroads in operation, and several companies undoubtedly enjoyed relatively long periods of prosperity, but trouble always seemed to be around the next bend. The problems came in varied forms—safety issues, worker dissatisfaction and unrest, the need for further investment to improve services and cope with new technology, the whims of hostile governments, and, perhaps most crucially, the arrival of new methods of transportation, such as the car, the semitruck, and later the airplane.

Initially, however, the railroads had a crucial advantage: for nearly the whole of the first 100 years following the opening of the Liverpool to Manchester line in 1830 (see pp.25–29) they were the only feasible form of transportation for many types of journeys. Thanks to the railroads, passengers could travel across whole continents, journey between a nation's major cities, commute between town centers and suburbs, and even reach remote villages—and freight, too, could be moved swiftly over great distances. By the beginning of the twentieth century, the railroads had become a sophisticated industry—larger and more influential than any other in existence at the time—and in the years leading up to World War I in 1914 they reached the apogee of their power. It was a time when the car was still the province of the rich and the lorry was an unreliable contraption, and both had to cope with roads that, for the most part, were rutted muddy tracks. It was a brief heyday for the railways, but it was one that had deep and lasting effects.

By 1914, virtually every country in the world had entered the railroad age. There were no absolute boundaries to the spread of the railroads, and even the toughest natural obstacles—jungles, mountains, rivers, and deserts—could be overcome by clever engineers. Latecomers, ranging from Costa Rica (1890) to Hong Kong (1910) and Morocco (1911), joined the established railroad nations in Europe and the Americas, extending the iron road across most of the globe. The United States, Europe, and Asia could boast transcontinental lines. Even some small islands had substantial systems. Sicily, the biggest island in the Mediterranean Sea, had more than 1,500 miles (2,400km) of line at the peak of its railroad age, and even the Isle of Wight, measuring just 150sq miles (380sq km), boasted 55 miles (89km) of line by the turn of the 20th century. In the Caribbean, the island of Cuba had 64 miles (103.5km) of railroad by 1849, mostly for carrying sugar.

Most countries embraced the railroads, particularly as their spread seemed so inevitable, but there were some exceptions. China was the last major nation in the world to give in to the incursion of the iron road. Even when the first line was finally built between Shanghai and Woosung in 1877, opposition to it was so strong, partly because it had been financed by foreign interests, that it was dismantled a year later. Gradually the powerful Chinese mandarins, or administrators, were persuaded of the necessity of joining the railroad age, although by 1895 a mere 18 miles (30km) had been completed. In contrast, today China has more high-speed rail lines than any other country (see pp.372–81).

For the most part, railroads played a key role in connecting the world in the 19th and early 20th century. Thanks to their facility for carrying both passengers and freight in large numbers, railroads began the process of globalization that was carried forward in the late

CROSSING THE WILDERNESS
Finished in 1890, the Costa Rican railroad traversed dense jungles, mountainous terrain, and raging rivers. The line was built to make the export of the country's main export—coffee—more efficient.

RULING THE RAILROADS
During the Mexican Revolution (1910–20), the railroads had great strategic importance, and whoever controlled the tracks, controlled the country.

20th and early 21st centuries by airplanes and information technology. In the latter years of the 19th century and the start of the 20th, railroads across the world grew on average by 10,000 miles (16,000km) per year, making many towns and villages accessible to the outside world for the first time.

The precise impact of the railroad varied across the world, but it was invariably profound. Unlike roads, which need little day-to-day attention, railroads require constant maintenance, such as patrols that ensure the good condition of the track, and investment, such as replacement of rails and signaling equipment. Therefore, once the railroads arrived, they transformed the economy and, inevitably, the character of a region. In fact, the railroads were a revolutionary force in both predictable and unpredictable ways. The most obvious advantage was a reduction in the cost of transportation. Consequently local produce, whether it was crops, minerals, or manufactured goods, could be transported more cheaply to national or global markets. The mail-order industry grew hugely at this time, as the railroads transported all sorts of mail-order goods to the newly connected citizens. The railroads also stimulated international population movement: immigrants arrived in the US by ship but then transferred to trains to fan out across the country. Indeed, many of the tracks were built especially to transport the influx of workers to industrial or agricultural centers, and then to transport away the results of their labors. Within countries too, the railroads were a catalyst for the vast migration of people—the towns and cities became attractions for people from the countryside who could now relocate far more easily.

British economist Alfred Marshall, writing in 1890, summed up the influence of the railroads and the industrial boom they had stimulated: "the dominant economic fact of our age is the development not of the manufacturing but of the transport industries." Another important outcome of the railroad boom concerned the workers and industries that supported the railroads themselves. The railroads required a whole set of new skills to enable them to run very large enterprises so, according to the railroad historian Terry Gourvish, "it is not an exaggeration to say that the [rail] industry played a key role in encouraging the growth of occupational professionalism based on specialized work. Engineering, law, accountancy, and surveying all received an important stimulus." Banks developed new loan systems in order to provide investment capital, universities stepped up to

BY 1914 THE IRON ROAD SPANNED

750,000 MILES
(1,200,000 KM)

supply competent engineers and surveyors, and factories of all kinds were built to manufacture the vast array of equipment needed, ranging from huge steel components such as boilers and wheels for locomotives to soft furnishings for seats and panels for train roofs.

Furthermore, the railroads affected other industries. By making transportation cheaper, they enabled similar factories to be concentrated in particular areas, which enabled the easy transfer of skills and experienced workers. The railroads also stimulated small-time capitalism, empowering many people previously restricted by their geographical isolation. In Mexico, Teresa Miriam van Hoy, the author of a social history of the railroads, found that the railroads introduced local competition in more remote regions since they:

> prompted the arrival of multiple suppliers, thereby breaking any monopolies or market strangleholds, and provided smallholders affordable access to markets beyond their local community.

In Russia, the village money-lender became redundant because the local peasants were now able to travel to the town market by train to sell their produce and turn it into cash. The new stations, according to one contemporary writer:

> swarmed with a mass of small traders, exporters, and commission merchants, all buying grain, hemp, hides, lard, sheepskin, down, and bristles—in a word everything bound for either the domestic or the foreign market.

Many of the lines built, especially in the latter stages of the railroad boom, were unprofitable but nevertheless had a lasting effect on the region they served. A railroad built in Senegal in 1885 as a way of establishing French colonial rule became a vital lifeline for the economy as, according to one historian of the African railroads, it allowed "the rubber, the cereals, and the peanuts from a rich hinterland to reach the Senegal river and transported [back to] the interior manufactures produced on the coast, such as textiles, foodstuffs and machinery." This story of economic opportunity was replicated across the world.

The railroads not only revolutionized existing industries, they also helped to create new ones: in the US, Birmingham in Alabama was a sleepy backwater until it was transformed into an industrial center by the Louisville and Nashville Railroad, which provided favorable freight rates, thus enabling the iron ore deposits at nearby Red Mountain to be exploited. The wine industry also developed, and not solely because of the cheaper transportation. In Argentina, wine production centered on the inland town of Mendoza and, as European immigrants arrived by train, they modernized the small existing vineyards and then exported their vastly increased yield via the railroads. Italian favorite Chianti became a regular feature of French and British restaurant tables thanks to quicker, easier access to wider markets. In other regions the taste of

TRANSPORTING WHEAT IN DAKAR
The railroads revolutionized modern commerce.
Dakar station in Senegal became a thriving center
of trade, transporting wheat and other commodities.

wine improved notably thanks to the railroads. As the rail historian and wine writer Nicholas Faith recounts, "in pre-railway days, many wines tasted decidedly resinous because they had been carried on mule-back in hog skins painted with pitch."

It was not only industries and economies that were affected by the railroads. Even the lives of those who could not afford train fares were improved by the railroad's existence. In many countries, particularly in South America, Asia, and Africa, the railroad provided the only safe thoroughfare for pedestrians—provided they did not get in the way of the trains, of course. The railroad lines forded rivers and canyons and cut through mountains far more efficiently than the old mule paths that were circuitous and badly maintained. Pipelines also followed many railroad routes, bringing water to many towns and villages for the first time. Even the station buildings became

ICE TRAIN
One strange industry facilitated by the railroads was the transportation of ice in the US. Until the advent of refrigeration, natural ice was harvested in cool northern regions and transported via railroad to the warmer south.

prominent local landmarks. They were often the most imposing building in the area, or even the only permanent ones, and they were frequently used as community meeting places. Every railroad also had a telegraph system, which allowed faster communication than ever before. Finally, thanks to the iron road, people were free to travel around spreading ideas, and information and newspapers could be distributed easily. Thus, the wider dissemination of democracy, and other political and social ideas, can be attributed—at least in part—to the railroads.

By 1914, in many countries, the railroad network was virtually complete. Consequently, the railroad companies were able to focus investment on improvements, such as faster locomotives or straighter track. Moreover, they could also devote substantial resources to making life more comfortable for passengers, especially those at the luxury end (see pp.170–77) of the market. These included elegant dining cars, comfortable sleeping facilities, and luxurious waiting rooms. At the other end of the scale, however, there were still lots of lousy trains. A branch line shuttle might run only a couple of times a day, and was not only slow but also subject to regular delays—the passenger cars might be pulled next to freight cars, which could be switched out at various stops along the way, slowing progress. The local trains that meandered between mainline towns often used the oldest rolling stock and made their way in a desultory fashion, waiting patiently on sidings for express trains to pass. Timetables were also subject to all kinds of vagaries, and were often designed for the convenience of the railroad company rather than its passengers. The railroads were unchallenged and the companies took full advantage, with many making healthy profits by charging high fares and bullying their smaller rivals to improve their own position. Worst of all was the grime—steam locomotives were dirty machines, spewing smoke and grit wherever they went.

So, while for a short period the railroads were king, their dominance could not last. By 1914, the railroads had done their job as the catalyst for the creation of the modern world. The world would never be the same again, but nor would the railroads. By the end of the World War I in 1918, semitrucks had become more sophisticated and would soon offer viable alternatives to rail freight transportation. Moreover, in the coming years the automobile would become ubiquitous, further reducing the dominance of the passenger railroad.

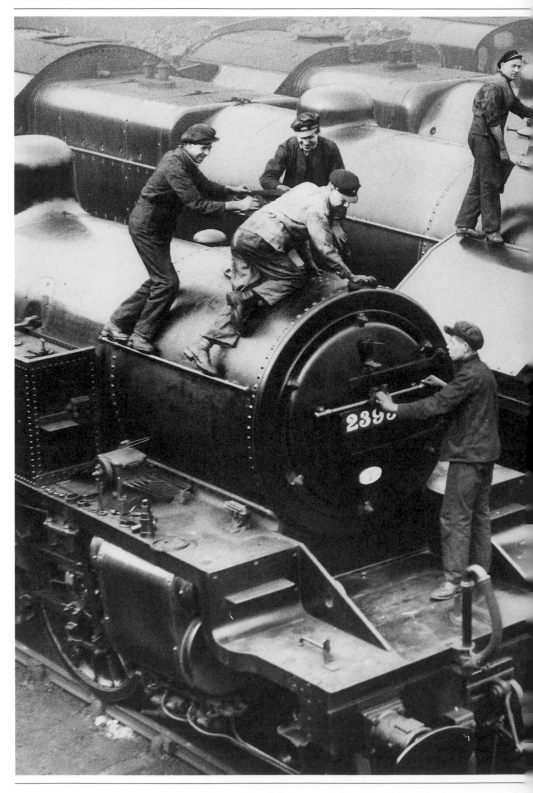

WORKING IN THE DEPOT
Workers at the Longsight locomotive sheds in
Manchester, England, polish a set of locomotives
in preparation for the Whitsun holiday rush of
1936. At the time, the railroads were still the most
popular means of long-distance travel.

The Field Railroads of World War I

A S STRATEGIC ASSETS, railroads came of age during World War I. At the start of the war, they were used chiefly to transport troops as close to the front as possible, after which soldiers still had a long march to their posts, usually burdened with supplies and equipment. However, as the two sides fought themselves to a stalemate on the Western Front, smaller, narrow-gauge lines began to proliferate locally, serving as connections between the main lines and the trenches. The Germans, possibly having anticipated the stalemate, were better prepared than the Allies. They had stockpiled huge quantities of 2ft- (60cm-) gauge rail track, having devised the concept during a successful but bloody campaign to colonize South West Africa (now Namibia) in a three-year war that began in 1904. The *Feldbahn*—literally "field railways"—were very flexible, and could be laid very quickly to transport troops across the vast Namibian plains.

The *Feldbahn* trains carried both troops and supplies, and were powered by little steam—or even gasoline-driven—locomotives which, by virtue of having eight wheels, the leading and trailing pairs of which could swivel and move sideways, could cope with short-radius curves of track. They could also be hauled by horses, of which there were plenty in each infantry unit. When they invaded Belgium and France, the Germans took enough 2ft- (60cm-) gauge railroad equipment to lay several hundred miles of track. The French, too, were well prepared. They regarded themselves as the inventors of the mini-railroad, having used it extensively in their invasion of Morocco in 1911. And so, as soon as trenches began to be dug in 1914, the French brought some 400 miles (645km) of narrow-gauge track (manufactured by Decauville) out of storage. The standard French narrow-gauge locomotive was the Pechot double-truck, double-boiler, eight-wheeler, but this was soon supplemented by vast numbers of small, saddle-tank locomotives from the great US engine manufacturer Baldwin.

The Russians—despite the many inadequacies of their army and their preparation for war—had also anticipated the need for railroads, having used them in their war with Japan in 1904–05. In that conflict, they had built a 30-mile (48-km), horse-drawn, narrow-gauge line

HEADING FOR THE EASTERN FRONT
German troops travel by train to the Eastern Front in 1914.
Large troop-carriers such as this took the soldiers to the
railhead, from which light railroads took them to the front.

near Mukden, which had greatly facilitated troop deployment. In
1914, the Russian army had nine railroad battallions, of which three
used narrow gauge tracks. Once the Eastern Front, which was longer
and more fluid than the Western Front, was established, a staggering
number of lines were built. In addition to some 560 miles (900km) of
preexisting track, a further 2,500 miles (4,000km) were laid, about
half of which was used by horse-hauled trains, and most of the rest by
steam engines. The Russians had the most need for this flexible
railroad system, since they were fighting both the Germans and the
Austrians across a wide swathe of Eastern Europe. The Austrians also
had a well-developed field-railroad system, especially for their
campaign in the south against Italy. In the mountains of the
Dolomites, they added to the existing narrow-gauge lines—which
were, in fact, slightly bigger than the standard 2ft (60cm), as they used
a 2½ft- (75cm-) gauge—to keep their mountain troops supplied. As
historian John Westwood writes in *Railways at War*:

> Anticipating war in the Dolomites, the Austrians had realized that it
> would be hard to build standard-gauge lines on the Italian–Austrian
> front, so a proportionately heavier task had been allocated to the narrow-
> gauge lines, which were expected to be quite long. Indeed, in the long

and usually fairly static campaign against Italy from 1915 to 1917, three substantial narrow-gauge lines were built by the Austrian army: the line from Auer to Predazzo, for instance, was thirty miles long, and included six tunnels and fourteen big bridges.

All of this was a far cry from the attitude of the British. British military strategists had expected a fluid war of movement, with troops attacking and counter-attacking each other across large areas, and so they were ill prepared when the Western Front became entrenched. According to a report into the British Army's use of transportation during the war, the authorities could not believe that the stalemate would continue:

> For the first two years of the war, the British transport arrangements were dominated by the idea that the war would soon revert to one of movement, that it was useless to embark on any large scheme which might be left far in the rear and become valueless before it had materialized, and become of use.

VITAL SUPPLY LINE
Allied troops lay a light railroad on the Western Front in 1918. The line brought supplies to the front and enabled troops to move swiftly between the trenches.

As a result, the British Army devoted more energy on trying to harness road transportation for the war effort than either their allies or their enemies. It was a hopeless task—what roads there were in rural France soon became impassable, and very often there were none to the front line. A British official report described the situation surprisingly eloquently:

> ... beyond the roads lay a nightmare quagmire of pulverized fields, ruined ditches, and flooded shell holes, threaded by temporary duckboard tracks and communications trenches. Through this muddy wasteland, every single item needed by the troops—food, water, clothing, medical supplies, tools, timber, barbed wire, mortars, machine guns, rifles, ammunition, and yet more ammunition—had to be carried.

It was an unedifying and perilous task. Hundreds of men died as they wandered off the duckboards in the dark and into flooded shell craters, where they were weighed down by their huge backpacks and drowned.

Narrow-gauge railroad lines were the perfect answer to this logistical problem. Their main advantage was that in the precarious conditions of the front line they were more flexible and efficient than any other form of transportation. They could be laid easily, with a minimum of ballast and only basic ties, and were also very easy to repair if they were shelled or damaged by the constant traffic they had to carry. As the name "field railway" implies, they could easily be lifted up and used elsewhere if the front line moved. The British, seeing the French success in using light railroads, belatedly began to develop their own from the summer of 1915.

The initial British light railroads were crude affairs that were mostly man-hauled, although mules were occasionally used. However, the beasts were at times reluctant to do their job, especially at night, when most of the operations were carried out. Far more sophisticated networks were later developed, and gradually two different types of narrow-gauge railroad emerged. The first type mainly ran from railheads to depots near the front line, and was normally worked by gasoline or gasoline-electric locomotives, or even steam engines in some cases. The second type of line—sometimes even narrower than 2-ft (60-cm) gauge, which made it seem almost like a toy train track—was a cruder design that reached right up to the trenches. Often called "tramways" by the British, these lines mostly used men or mules for traction, partly because they were so close to the enemy that the noise of the engines might attract attention and possible shellfire.

The two systems were supposed to be kept independent of each other, since the lines nearer the front were not sufficiently robust to carry the larger loads used by the lines from the railhead. All the lines were necessarily short, usually between 5 and 15 miles (8 and 24km) long, and required almost constant maintenance. Derailments were common, particularly when tanks and artillery were being transported, and were usually dealt with by manpower alone—a few men would be called upon to heave the engines or cars back onto the tracks.

Trains were forced to operate under the cover of darkness, and the only light—if one were used at all, which was impossible near the front—was the size of a small flashlight. Yet nearly all the lines were single-track, and there was no signaling system, so operations were carried out on a "line of sight" system. This meant that the engineers had to stay constantly alert, checking whether there was a train ahead that had unexpectedly halted or broken down. A telegraph system could be used to contact the rail traffic controller, but only if there were problems. The unsophisticated nature of the system is best revealed by the fact that the engineers often had to resort to finding water for their steam locomotives from the nearest shell hole, as there was often no other water source. One harebrained idea inspired by the lack of rolling stock was to adapt Model T Ford automobiles. They were fixed to a rail chassis, but proved too light for the task—they slipped on the rails due to insufficient adhesion, so the idea was abandoned.

Given the poor state of the lines and the frequent use of men or mules, speeds were very slow. Nevertheless, these little toy-town railroads were infinitely better than any other form of transportation at the front. There were, though, limitations. The maximum load of a narrow-gauge train was 30 tons (33.5 tonnes) and consequently, when

"The Russians had perhaps the greatest need of such lines… given their exceptionally long front line"

JOHN WESTWOOD, *RAILWAYS AT WAR*

MULE TRAIN
French soldiers of the 11th Artillery
Regiment complete a 2-ft (60-cm)
narrow-gauge railroad near
Soissons, France, in 1917. Mules
stand ready to haul the cars.

supplies were transferred at the railhead, at least ten such trains were required to take the load from one mainline service. Nevertheless, the lines carried huge loads. An officer of the railroad corps reckoned that one line could carry up to 1,200 tons (1,340 tonnes) of materials each night, representing up to 150 trains each run by a couple of men. Even large guns were carried, often straddling more than one car. Toward the end of the war, a new use was found for the lines: field guns were set up on cars, enabling the gun to be moved after a few shots were fired, thus preventing the enemy from locating them. And it was not only ammunition and supplies that were carried—whenever possible, the trains carried troops to and from the front, saving the men hours of trudging through heavy mud where, in darkness, they risked falling into shell holes.

John Westwood concludes that "all the belligerents, even the Germans, made greater use of the narrow-gauge railways than they had expected." This was partly as a result of the stalemate that lasted for three and a half years on the Western Front, but also because of the state of transportation technology at the time. Field trains were an ideal solution for the logistical problems of the war's muddy battlefields, and they became ubiquitous. It was only when the Germans, at last, broke through the lines in the spring of 1918 and the Allies then counter-attacked that the lines lost their purpose. Because of their temporary nature, very few survive today, with just a handful of sections of line in northern France preserved to show what *Les Petits Trains*, as the French call them, achieved in that terrible conflict.

American Luxury

Passenger services began in the US in the 1830s, and by 1869 the first journeys across the whole continent were made. Long-distance travel led to innovations such as Pullman sleeper cars (see pp.170–77) and observatory cars.

WM NO.203 (1914)
A luxury Pullman sleeping car used by the president of the Western Maryland Railroad, No.203 had a steel frame with wood-effect exterior styling. It had an observation lounge, a dining area, and four sleeping berths.

RDG NO.800 (1931)
No.800 was the first electric multiple unit (EMU) to be put into service by the Reading Company on its Pennsylvania commuter routes. An EMU train consists of self-propelled cars—No.800 was powered by 11,000 volts of AC current, collected via a pantograph from overhead lines.

NW NO.512:
POWHATAN ARROW (1949)
No.512 was a 51-class lightweight steel car built by Pullman. It was one of the units that made up the *Powhatan Arrow*, Norfolk and Western's streamliner passenger service that served a 676-mile (1,088km) route from Norfolk, Virginia, to Cincinnati, Ohio.

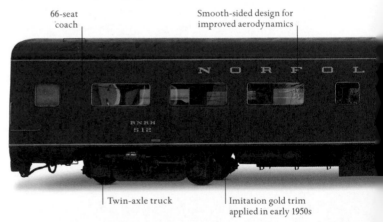

66-seat coach

Smooth-sided design for improved aerodynamics

Twin-axle truck

Imitation gold trim applied in early 1950s

NW NO.1489 *SCIOTO COUNTY* (1949)
Named after a county in Ohio, *Scioto County* was built by Budd Company for the Norfolk and Western Railroad as a sleeper equipped with 10 roomettes and six double-bedrooms. After several refits and spells as a commuter car and snack-bar coach, it was retired in 2001 before returning to service on a heritage line.

BOMX NO.130: *HERSHEY WARE* (1949)
Named *Hershey Ware* by the Baltimore and
Ohio Railroad Museum after a restoration,
No.130 was a passenger car built by Budd

Company in Philadelphia. Originally used
as a 21-berth sleeper by the Pennsylvania
Railroad, it was refitted as a commuter car
in 1963 for the New York World's Fair.

RDG OBSERVATION NO.1 (1937)
This Budd Company observation car was
the rear car in Reading Company's flagship
Crusader train, a streamliner service that

linked Jersey City and Philadelphia. It
boasted ultra-modern features including
air conditioning, sound proofing, and
movable arm chair seating.

Original Tuscan red
and black livery

Hauled by NW J-class
locomotive (see p.293)

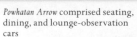

Powhatan Arrow comprised seating,
dining, and lounge-observation
cars

BALTIMORE AND OHIO NO.1961 (1956)
A Budd Company self-propelled dining car,
No.1961 was powered by two underfloor
diesel engines. The rear of the car could seat

24 regular passengers, while the eight-table
dining area was served by a full kitchen. It
was converted for conventional passenger
duties in 1963 before being retired in 1984.

Wartime Railroad Disasters

IT IS NO COINCIDENCE THAT A HIGH PROPORTION of the world's most serious rail disasters occurred during wartime. Britain, France, and Italy all experienced their worst accidents in terms of fatalities during the two world wars—although interestingly, none involved enemy attack—and the highest death toll of any rail accident in Europe occurred in Romania during World War I. Each of these disasters was wholly or partly caused by the overuse of the railroads due to the imperatives of war, combined with a general decrease in safety. Wartime censorship meant that information on these incidents was withheld at the time, and even today the details are sketchy. Consequently, many wartime disasters have been largely forgotten.

The first of this series of tragedies occurred during World War I at Quintinshill, near the English–Scottish border. In terms of loss of life, it remains by far the worst train accident in British history, and

while the direct cause was a series of mistakes by rail traffic controllers, a contributory factor was the enormous pressure placed on the railroads due to the war. The Caledonian main line approaching Carlisle from the north—one of two main rail connections between England and Scotland—was one of the busiest stretches of railroad in the country during the conflict. A huge number of "Jellicoe specials"—freight trains carrying coal for Admiral Jellicoe's Royal Navy—used the line when returning empty from Scotland to England, as well as local and express passenger services.

On the morning of May 22, 1915, the two overnight passenger sleeper expresses from London were late, as often happened in the war due to the intensity of traffic. The small interlocking towers at Quintinshill, 10 miles (16km) north of Carlisle, controlled a section of the main line as well as the sidings on either side of the track, which were used to temporarily accommodate slower freight trains or local services so that faster trains could pass. That morning, the sidings were full of empty Jellicoe specials waiting to return to the mines to collect more coal. Consequently, the

HEADING FOR THE FRONT
During World War I, trains were used on an unprecedented scale to transport troops and equipment to the battlefield. Here, French soldiers depart for the Western Front in 1914.

TRIPLE COLLISION AT QUINTINSHILL
An express engine lies in the wreckage of an earlier collision between a local service and a troop train at Quintinshill, England, in 1915. A freight wagon (left) has been thrown off the track.

towermen decided to direct a slow local train off the northbound main line and onto the southbound track, in order to allow the passenger sleeper express trains to pass through.

Disastrously, the rail traffic controllers then forgot that the southbound track was occupied, even though the parked train was within sight of the interlocking tower. It was just after 6am, and the signalmen were about to change shifts—George Meakin giving way to James Tinsley. However, against the rules, the two men had agreed to swap shifts a little later than scheduled to give Tinsley time to take the local train to the interlocking tower. Tinsley, therefore, was busy filling in the register—to cover the fact that he had not started work until after 6am—and chatting about the war when he gave the signal "line clear" to a southbound troop train. He had forgotten that the local train—on which he had just traveled—was sitting on the southbound main line.

The troop train was carrying 485 soldiers of the Royal Scots, who had just finished their training and were bound for the fighting at Gallipoli in Turkey. The engineer had no chance of stopping when the troop train came down a slight grade at more than 70mph (110kph). The train smashed into the local service head on, with such force that the troop train's cars were telescoped into a length of just 210ft (64m)— a third of their original size. To compound the disaster, the old rolling stock that had been commandeered for the troop train was made of wood and acted like a tinderbox—a fire quickly broke out and was fueled by kerosene lanterns and coal from the engines. But worse was yet to come. The northbound express for which the local service had been waiting was unable to stop, and plowed into the wreckage from the two earlier trains, which had been strewn across the northbound main line. The death toll from the incident was 227—by far the worst British rail disaster, and nearly twice the total of the second-worst, at Harrow in 1952—but fatalities on the two passenger trains were light, although the figures may have been massaged by the wartime official sources. The two towermen were jailed for manslaughter, with relatively light sentences given the scale of the disaster—Tinsley received three years and Meakin half that.

The Romanian accident, which happened on January 13, 1917, is shrouded in mystery because of its location and the tight censorship of the Romanian and Russian authorities. It bore several similarities with a later disaster at St-Michel-de-Maurienne in France (see p.285)—a

TRAGEDY AT CIUREA, 1917
It is thought that the devastating crash was partly
caused by soldiers in the overcrowded cars
accidentally damaging brake pipes.

heavily overloaded troop train ran out of control, leading to a fire that
killed many of the victims. The accident occurred at Ciurea—in a
remote eastern part of Romania near what is now the border with
Belarus—and involved Russian troops and Romanian civilians fleeing
a brutal German advance. Romania had entered the war late on the
side of the Allies, and after early success was soon overrun by German
forces. To escape the enemy, a huge train of 26 cars packed with
wounded Russian soldiers, as well as refugees, left the small town of
Bârnova bound for Ciurea. A survivor, Nicolae Dunanreanu, wrote
of the scramble to get on the train:

> ... everywhere, people—and particularly soldiers—clambered on to
> the roofs, steps, and buffers, gripping each other in mad desperation.
> There was not even the smallest corner free, one could not even get both
> feet on a step, nor a buffer, and these desperate people seeking a relative
> or fleeing from the enemy who occupied more than half the country
> could not guess that a greater disaster awaited them.

The two stations were separated by an incline that averaged 1 in 40,
but with sections as steep as 1 in 15. It became apparent shortly after
starting the descent that the train's brakes were not working
properly—it later emerged that passengers had broken the connecting

DISASTER IN THE ALPS
The remains of a 19-car train carrying over 900
French troops lie at the bottom of a gorge near the
village of St-Michel-de-Maurienne, France.

pipes between cars by stepping on them as they crowded onto the
train. The two locomotives did not have sufficient braking power
between them, and consequently the train hurtled ever faster down
the slope. Despite the efforts of the train crew—who took the
emergency measure of putting the locomotive in reverse, and tried to
sand the track to increase adhesion—the cars were derailed as they
entered Ciurea station, causing destruction on a vast scale. The final
death toll is thought to have exceeded 1,000, although wartime
secrecy—and the remoteness of the area in which the accident
occurred—meant that no precise figure has ever been ascertained.
There is no doubt, however, that it was by far the worst railroad
accident ever to occur in Europe as a whole.

Later that year, an accident occurred in France that would prove
to be the worst railroad disaster within Western Europe. It was
caused by elementary mistakes made by railroad officials working
under the strain of wartime loads and—crucially—under the
orders of the military, who ignored the officials' warnings. A very
long train of 19 cars was being hauled over the Alps on the night of
December 12, 1917, carrying more than 900 French troops on their
way home for Christmas. The men had fought in Italy and were

anxious to get home quickly for their leave. Having traveled through the Mont Cenis tunnel, a crucial link between the two countries, the train waited at Modane on the French side for more than an hour as other services were allowed onto the overburdened line. The train was also being delayed because of the lack of a second locomotive, which was vital not just to provide extra power up the grades, but to assist with braking on the descents. Only three cars had air brakes, while the rest had either crude, hand-operated brakes, or none at all.

As the delay lengthened and the *poilus* (French infantry) became rowdy, the train's engineer, Girard, came under pressure to proceed down the incline. He refused unless a second locomotive could be found, but the only one available had been allocated to an ammunition train. Girard was overruled by the local military traffic officer, Capitaine Fayolle, who told the engineer that he would be thrown into the *fortresse* (prison) if he refused. It was a classic case of military personnel failing to understand the limitations and safety requirements of the railroad. As a result, the inevitable happened. With such a huge load and inadequate braking power, the train began to speed out of control. When the brakes were applied, the friction was so great that they heated up and caught fire. This sowed panic among the passengers, some of whom jumped off the speeding train. Traveling at around three times the speed limit of 25mph (40kph), the train jumped the rails at a bend near the village of St-Michel-de-Maurienne. Several cars plunged into the gorge below, while others burst into flames. Relieved of its burden, the locomotive stayed on the tracks—and Girard, who was not initially aware of having lost his load, survived.

The death toll was initially announced as 424, but is now thought to have been 457. Other estimates put the number as high as 675, since many of the dead were incinerated in the ensuing fire—which took a day to burn out—and several survivors later succumbed to their injuries in the hospital. But it could have been even worse. Only the quick actions of a stationmaster

NUMBER OF UNIDENTIFIED DEAD AT ST-MICHEL-DE-MAURIENNE

135

prevented a train carrying Scottish troops toward Italy from crashing into the debris. Reflecting the sensitivity of such wartime incidents, it was not until 79 years after the accident, on December 12, 1996, that a memorial to the dead was opened at the site of the disaster.

Spain experienced its worst train disaster during World War II, despite the country's neutrality. The accident occurred on January 3, 1944, near the village of Torre del Bierzo in the León province, when three trains collided inside a tunnel. Like the Romanian and French disasters, the cause was a runaway train speeding down an incline, resulting in a fire that claimed most of the lives. The overnight Galician mail express failed to make a scheduled stop at Albares due to a broken braking system. The stationmaster at Torre del Bierzo, the next station down the line, ordered railroad ties to be placed on the line to slow the train down, but his efforts were to no avail. The train ran toward a tunnel where it hit another train that was in the process of being moved out of its path. Unaware of the crash, a coal train with 27 loaded cars then approached the tunnel from the opposite direction and plowed into the wreckage. The ensuing fire burned for two days, preventing the injured from being rescued and making identification of most of the victims impossible.

Strict censorship under the regime of General Franco meant that the accident received very little publicity at the time, and the official RENFE (the Spanish rail company) file on the accident was lost. There were many illegal travelers on the train heading for a post-Christmas market, and although the official death toll was 78, research has shown that the real figure was closer to 500.

World War II was also the backdrop for Italy's most serious accident, which brought a death toll far in excess of any other rail disaster in the country. Again, wartime conditions were the underlying cause. The accident happened at Balvano, a small town inland from Salerno on the Bay of Naples. The area was under occupation by Anglo-American forces that had battled their way up from Sicily, and food and other basics were in short supply. Many townspeople jumped on freight trains illegally to travel into the countryside to obtain supplies, either for themselves or to sell on the black market. One such steam-hauled train left Salerno on the wet and cold evening of March 2, 1944, heading for farms inland in the Apennine mountains. Hundreds of people jumped onto the flat cars at each successive stop, and sheltered under tarpaulins and whatever else they could find. The train stopped in a tunnel near Balvano, where it was forced to wait nearly 40 minutes for another service to come down the hill. This was to prove fatal for hundreds of the 650 or so illegal travelers. Wartime shortages meant that the only fuel available for the engine was poor-quality coal, which emitted a high level of carbon monoxide. The incline in the tunnel caused the fumes to spread downward; the death toll was enormous, and has been estimated at between 450 and 500. Those who survived had mostly been in the rear cars, which were not in the tunnel when the train stopped. The alert was sounded by a brakeman, who ran back to the nearest station shouting "they are all dead," before collapsing from the effects of the fumes. A colonel in the US Army who helped in the aftermath of the disaster later wrote: "The faces of the victims were mostly peaceful. They showed no sign of suffering. Many were sitting upright or in positions they might assume while sleeping normally." Death had come quietly and quickly.

Fortunately, the scale of these disasters is unlikely to be repeated in the modern era. Train technology—such as advances in braking efficiency—and the enforcement of regulations mean that train travel is far safer now. Generally, trains are not as overcrowded as in previous years, and fire is rarely a hazard given that steam locomotives are no longer used. Moreover, cars are no longer made of wood, and are designed and built to be strong enough to withstand the forces of crashes—so even when accidents do occur, the survival rate tends to be higher. Nevertheless, with high-speed trains traveling at nearly 200mph (320kph) every day in Europe, the possibility of a tragedy on a large scale remains—as evidenced by the July 2013 accident at Santiago de Compostela in Spain, in which 79 people lost their lives.

The Hejaz Railway

R AILROADS CAME LATE TO THE MIDDLE EAST. By the end
of the 19th century, there were just a few lines operating in the
crumbling Ottoman Empire. Their military and political value had
been recognized by far-sighted rulers, however, and this was the
catalyst for the construction of a railway deep in the desert of
the Hejaz region of what is now Saudi Arabia—a line that would be
made famous in the West by Lawrence of Arabia.

As with many pioneering railroads, the idea for its construction was
mooted long before work commenced. German-American civil
engineer Charles Zimpel proposed a line from Damascus to the Red
Sea in 1864, and numerous similar lines were also suggested in the last
third of the 19th century. However, it was the proposal in 1897 by
Muhammad Insha Allah, an Indian Muslim teacher and journalist,
that attracted the attention of the Sultan of the Empire, Abdulhamid II.

Abdulhamid was the conservative leader of the failing Ottoman
Empire, ruling from Constantinople (now Istanbul). Soon after his
accession, in 1876, the empire had lost two fifths of its territory, including
European holdings such as Bulgaria and Greece, effectively ending
Ottoman influence in Europe. Following this loss, the Sultan resolved to
strengthen his hold over the remaining
Asian part of the empire, which included the
Arabian Peninsula. The political influence of
the empire was waning, but Abdulhamid
hoped to reinforce its religious significance.
The proposed railroad offered an opportunity
to consolidate his own position as Caliph,
leader of all Muslims: the new line would
ensure a fast, cheap way for pilgrims to travel
to Mecca. Funding for the railroad was
obtained largely through Muslim support
due to its religious significance.

T. E. LAWRENCE
British Army officer T. E. Lawrence became
familiar with the Middle East during his career
as an archaeologist. He led the Arab Revolt in
1916–18, which all but destroyed the Hejaz Railway.

THE HEJAZ RAILWAY

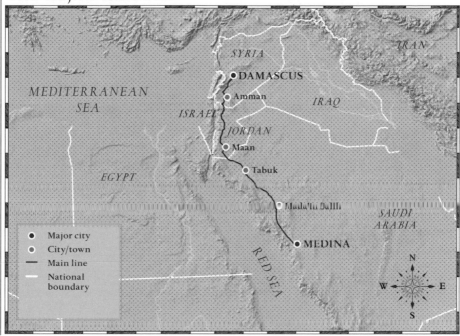

Work started on the 4ft 11¹¹⁄₁₆in- (1,500mm-) gauge railroad in late 1900 under the auspices of a German engineer, Heinrich Meissner, who drove the project for eight years. Initially, it was beset by problems and progress was slow. The original surveys were unsatisfactory and had to be redone. The laborers, mostly conscripts, worked in appalling conditions—poor treatment that resulted in a mutiny. Recognizing the lack of progress, the Sultan took a softer approach and Meissner began to impose a more acceptable regime, attracting experienced foreign workers from Belgium, France, and Germany in particular. In the later stages, however, Christians were not allowed to work on the southern end of the railroad because of religious sensibilities, but by then trained Turkish Muslim engineers were available.

Three teams were created: for reconnoitring, surveying, and construction. The reconnoitring team, which made a preliminary assessment of the route, experienced the most difficulties: they went into the desert mounted on camels and horses, venturing into unmapped land and facing hostile tribes, and so were accompanied by a cavalry detachment. The second team, the surveyors, used the maps created by the reconnoitring group to set out a detailed route for the

"The engineers had to build a railway through this precipice, the 'belly of the demon'"

DOMINICAN FRIAR, TRAVELING ON THE HEJAZ RAILWAY IN 1909

construction gangs to follow. Construction was carried out by a special railroad battalion with four divisions, each one focused on a particular task: the advance party marked out a trace for the track and prepared the earthworks; the second division put down the ballast; the ties were laid by the third group; and the rails were laid in by the fourth.

Such a disciplined and organized process allowed rapid progress despite the difficulties of the terrain and other obstacles. It was not just the heat, which could reach 122°F (50°C) in the middle of the day (matched by cold nights in winter), or the remoteness of the land, but the sheer scale of the operation: 1,000 miles (1,600km) of track had to be laid to reach the original target—the holy city of Islam, Mecca.

The scarcity of water was the worst problem faced by the builders of the line, which followed old pilgrimage trails. These had occasional wells and pools where rainwater collected, but, for the most part, the vital liquid was stored in cisterns installed along the line and replenished by tank cars transported down the newly constructed railroad. While lack of water was a perennial problem, ironically flash floods washed away parts of the line in the rainy season. To limit such damage, many sections of the railroad were built on embankments.

Sand drifts across the tracks were another obstacle. When construction reached the desert of the southern Arabian Peninsula, there was little vegetation to prevent sand being driven onto the railroad by the wind, and it was therefore necessary for the line to be protected by sandbanks made of stone and clay.

The greatest irony was the shortage of fuel. Coal mined in Turkey proved too smoky, and so fuel was imported from Wales at great expense, then mixed with the local coal. Since steam locomotives can run on oil, the solution was in fact near at hand. As James Nicholson, author of the line's history, puts it, "In view of the fact that the Ottomans at that time also controlled the Eastern Province of what is now Saudi Arabia, they would have been surprised to discover just how easily all their fuel needs could have been met by what lay beneath the sands."

The workers, who numbered 7,000 at the peak of construction, lived in small tents that had to be moved forward constantly as the line progressed. Laborers lived on a diet of bread, cookies, or rice with only occasional additions of meat. There were no fresh vegetables or fruit, so vitamin deficiency diseases, such as scurvy, were common. Cholera outbreaks were not unusual and caused widespread panic, with workers fleeing the camps, which delayed progress.

As the railroad approached Medina, plans were still in place to extend the line to Mecca. However, opposition to what was called "the iron donkey" was strong among the local tribes, who did not want to see it reach Mecca. Their objections were not solely stimulated by religion: many local tribesmen made their living from operating the camel caravans for pilgrims. This resistance culminated in a revolt in January 1908, when Abdulhamid's political position was already weak. As a result the planned extension of the line to Mecca was scrapped. Instead, Medina, the second holiest city in Islam, became its terminus.

The last stretch of track from Al-'Ula to Medina was the most difficult to lay in terms of the terrain, but work speeded up with the arrival of extra workers. The rapid conclusion of the project was partly

FIRST TRAIN FOR MEDINA
A crowd gathers to send off the first train carrying passengers on the Hejaz Railway from Damascus to Medina on the pilgrimage to Mecca in 1909.

to ensure that the line's opening coincided with the anniversary of the Sultan's accession to the throne on September 1, 1908. The deadline was met and there were celebrations in Medina, but the Sultan did not attend: his popularity was by this time so low that he feared his absence from Constantinople might result in a coup.

Despite the Sultan's mounting political difficulties, the Hejaz Railway enjoyed eight years of normal operations, carrying many thousands of pilgrims to and from Medina. However, the line's fate was always bound up with regional and, indeed, global political considerations, and World War I was to turn the railroad into a battleground.

The Ottoman Empire entered the war in 1914 on the side of the Germans. The British were keen to ensure that the Turks did not launch attacks elsewhere, and so encouraged the Arabs in the Peninsula to rise against them. The Hejaz Railway was an obvious target and in June 1916 the tribes began attacking the line. However, the Arabs needed explosives and better equipment, and that is where T.E. Lawrence—Lawrence of Arabia—entered the fray. Despite ranking only as captain and holding a desk job in Cairo, he persuaded his superiors to send him across the Suez Canal to support the uprising led by Prince Feisal, one of the sons of Sharif Hussein, the Emir of Mecca.

Feisal's irregular troops had already launched several successful attacks on the line when Lawrence joined them in early 1917. He led several more raids, both attacking trains and sabotaging the track. The strategy, an early use of guerrilla tactics, was not to close the line, but rather to tie up Turkish troops in its protection. Very few of Lawrence's troops were killed in these raids—depicted so powerfully in the David Lean film *Lawrence of Arabia*—but thousands of Turks lost their lives and the strategy proved successful. Gradually, Lawrence and Feisal worked their way up the line and gained control of the railroad as the Turks fled north. This meant that Medina, and the Turkish troops defending it, were cut off from the rest of the Ottoman forces.

Lawrence and Feisal eventually joined up with the British forces under General Allenby for a final assault on Damascus. Feisal's army, which included Lawrence, was given the task of cutting off the junction that led from the Syrian city of Daraa to the Mediterranean port of Haifa (now in Israel). The last, decisive attack on the Turks in Damascus in September 1918 was successful, but Medina was still occupied by Turkish troops who did not surrender until January 1919, arguably the last action of World War I.

LAWRENCE OF ARABIA AND THE ARAB REVOLT
Arab soldiers attacking a train on the Hejaz Railway in
the film *Lawrence of Arabia*. Despite the notoriety he gained
for his exploits, Lawrence insisted it was the Arabs' war.

As Lawrence recognized in his account of the battle over the Hejaz,
Seven Pillars of Wisdom, the Turks were courageous and became adept at
repairing the line. Following the war, and the division of the Ottoman
Empire into states under British or French control, the operation of
the line was split between the two European colonial powers. The
Allies and the retreating Turks had destroyed much of the southern
section of the railroad, but several parts remained open for traffic.

The best-used sections were freight services on the branch line
between Damascus and Haifa and, until the start of the Syrian conflict,
there was a passenger train from Damascus to Amman (the capital of
Jordan). Saudi Arabia is also building a new network of lines to cater
for the Haj and its freight needs, and there has even been talk of
reopening the entire Hejaz Railway, but this may prove too expensive,
and unnecessary now that many pilgrims travel to Mecca by air.

The eight years from 1908 to 1916 were to be the only time at which
regular services operated along the whole railroad. Even then, conditions
were uncomfortable for the pilgrims, as services were overcrowded and
slow. The journey, however, which took just over a day, was a vast
improvement on the old 40-day overland trip. It proved a brief heyday
for one of the world's most ambitious railroad construction projects.

Streamliners

In the age of aviation and the motor car, locomotive designers sought to lure passengers back to the railroads with a new generation of high-speed, modern designs. First applied to the marquee passenger expresses of the US, the term "streamliner" came to describe the steam, diesel, and electric locomotives that were sculpted for speed.

Aerodynamic styling designed by famous steam locomotive engineer Sir Nigel Gresley

LNER NO.4468
MALLARD (1937)
A Class A4 Pacific built by London and North Eastern Railways, No.4468 *Mallard* holds the world record for the fastest speed achieved under steam traction. On July 3, 1938, it hauled seven coaches at a speed of 125mph (202kph), thanks to its aerodynamic bodywork and highly efficient steam circuit.

Streamlined valances hide running gear

ETAT ZZY 24408 (1934)
Built by French car maker Bugatti, ZZY 24408 was one of several "Autorail Rapide" express rail-cars used in France. Despite a number of innovations—such as drum brakes, four gasoline engines, oil-damped suspension, and a central cupola for the engineer—the model was withdrawn in 1953 due to the expense of its fuel.

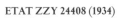

PPL NO.4094D (1939)
No.4094D was a "fireless" switcher run by Pennsylvania Power and Light Co. Instead of a firebox and boiler, it had a reservoir charged with steam from an external source. It was used in settings where pollution or fire risk had to be eliminated, such as gas power stations or chemical works.

NW NO.611 (1950)
With rigid wheelsets and lightweight driving rods to enable its relatively small driving wheels to reach 110mph (177kph), No.611 is a Norfolk and Western J-class steam engine. NW's flagship model, the class hauled both passenger and freight trains.

"Kylchap" smokestack contains four nozzles for improved engine efficiency

6ft 8in- (2m-) diameter driving wheels

DB CLASS 602 (1970)
Originally built in 1957 as a diesel-hydraulic VT 11.5 passenger train for the Trans-Europ Express, Deutsche Bundesbahn's Class 602 was a 1970 refit with 2,200-hp (1,600-kw) gas turbine engines. The new class was capable of 124mph (200kph), but high fuel costs led to its withdrawal in 1979.

HOKURIKU SHINKANSEN E7 (2013)
This model of Japanese *Shinkansen* bullet train (see pp.364–71) was launched in November 2013 in Rifu, Miyagi, Japan. These trains operate at speeds of around 200mph (320kph).

Australia's Gauge Bungle

THE CHECKERED HISTORY of railroads in Australia is in many respects a case study of how not to manage a railroad network. In contrast to many other countries, where the railroads acted as a great unifier and were built in a standardized way that assisted their expansion, the state-run railroad companies in Australia seemed to take an almost perverse enjoyment in making life difficult for freight carriers and people wishing to travel long distances across this vast country. In no other country in the world has the issue of the gauge of the various lines so dominated the history of the railroads and resulted in so much damage to the development of the rail network.

Given that in its infancy Australia was the destination of so many British convicts, deported there in the 19th century, it is perhaps unsurprising that its first railroad line was operated by convict-power. In 1836, an 5-mile (8-km) narrow-gauge line was built across the Tasman Peninsula in Tasmania to the Port Arthur prison settlement, enabling visitors to avoid a stormy sea passage. Convicts who had first been conscripted into building the track were then required to haul the little open-top cars that plied the line, enjoying a slight rest on downward sections, where they could hop aboard for the ride. Passengers paid the not-inconsiderable fare of one shilling. However, this human-powered arrangement could hardly be called a railroad, and it was not until 1854 that the first proper lines were opened.

Constructed almost simultaneously, the nation's first lines set a pattern of disconnected services that would plague Australia's railroad system throughout its history. In South Australia, an 7-mile (11-km) horse-drawn line opened in May 1854 between Goolwa on the lower Murray River and Port Elliot, and four months later the nation's first steam service began in Victoria, between Melbourne's Flinders Street Station and Sandbridge (now Port Melbourne). Both these lines used the 5ft 3in (1,600mm) broad gauge, as used in Ireland, but a line opened in New South Wales the following year using the 4ft 8½in (1,435mm) standard gauge. This was the start of Australia's failure to coordinate its railroads, resulting in—

THE NUMBER OF
DIFFERENT RAIL
GAUGES USED IN
AUSTRALIA

22

AUSTRALIA'S MAIN LINES

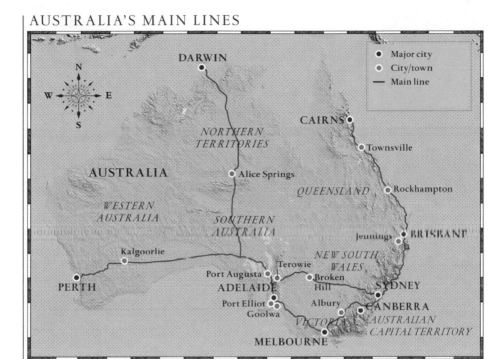

according to rail enthusiast and former deputy prime minister Tim Fischer—a veritable confusion of gauges, with 22 different track widths being used across the nation. Following the opening of these first lines, early attempts were made to coordinate the use of gauges. However, these only added to the confusion. Amazingly, the New South Wales line was first converted to the broader width and then back to standard gauge in the belief that this would match neighboring Victoria. However, Victoria and South Australia claimed that, in fact, changing to standard gauge would be too expensive, as they had bought rolling stock in broad gauge. And thus, as a historian of rail transportation in Australia put it, "the glorious bungle began." To exacerbate matters, the three remaining states, Queensland, Tasmania, and Western Australia—all of which were separate British colonies until the creation of the Commonwealth of Australia in 1901—each chose the 3ft 6in (1,067mm) narrow gauge, enabling cheaper construction but adding to the gauge confusion.

In the last quarter of the 19th century, the railroads began to expand rapidly, mostly due to the demand for freight conveyance. The overseas export market was the catalyst, with the consequent need to carry ore from the country's mines, and agricultural produce from the vast

BRIDGE ON THE GHAN RAILWAY
The Ghan transcontinental railway runs from Darwin
to Adelaide—a 1,851-mile (2,979-km) journey that has
only been possible since 2004, when the entire track
was finally converted from narrow to standard gauge.

interior, to Australia's ports. But any attempt to create a national
network—or even merely to create links across state borders—was
hampered by the gauge issue. In effect, the six Australian states each
built a railroad network resembling that of an individual country, with
very few connections and little cross-border traffic.

For example, when the rails of New South Wales and Victoria met at
Albury in 1883, in theory linking Melbourne and Sydney, the different
gauges meant that passengers had to change trains, and goods had to be
transhipped for onward dispatch. The same happened between Sydney
and Brisbane, when the lines between Queensland and New South
Wales met at Jennings in the same year. Even at borders between states
where there was no break of gauge, "huge time-wasting barriers were
created to stop any kind of seamlessness," according to Fischer. The
only two adjoining states with the same gauge—Victoria and South
Australia—changed locomotives at the border because each state had
bought different types of engine that could not fit onto the other state's
network. This lack of standardization proved costly, as it hindered

savings from bulk-buying of equipment. Coastal shipping took up the slack, handling most passenger and freight traffic between states, even though maritime transport was a slower method of travel.

The question of whether all the railroads should adopt standard gauge had been debated as early as 1890. However, it was not until Australia became an autonomous nation in 1901 that the gauge issue was addressed, by the new federal government. Before then, the various colonial administrations had decided that the cost of standardization would be too high, making differences in gauges—known as "breaks of gauge"—inevitable. In 1917, the government built the standard-gauge Trans-Australian Railway between Kalgoorlie in Western Australia and Port Augusta in South Australia, but several breaks of gauge were required to reach connecting stations—at Kalgoorlie to reach Perth, and at Port Augusta and Terowie to reach Adelaide. Similarly, when the line between Sydney and South Australia was later completed in 1927, breaks of gauge were required at Broken Hill, where the line switched to narrow gauge, and at Terowie to reach Adelaide.

The nationwide rail system had expanded to some 26,000 miles (40,000km)—relatively small for so large a country—by the end of World War I. As in other countries, railroads lost market share between the wars as they faced growing competition from road transportation, even in the key freight market. Starved of much-needed investment from their state owners, the railroads then had to cope with the added loads of wartime, following Australia's entry into World War II in September 1939. The railroads were used intensively for the war effort, carrying both troops and freight, but afterward a pattern of neglect began to threaten their very existence. Facing even greater post-war competition from road transportation, and with the emerging airlines picking off long-distance passengers, there was talk of abandoning the system. In the words of one historian of the Australian railroads: "they might have succumbed to the direst predictions of the critics and become nothing more than streaks of rust across the countryside…"

THE DISTANCE THE TRANS-AUSTRALIAN RAILWAY RUNS WITHOUT A BEND

297 MILES

(487 KM)

THE LONGEST STRAIGHT TRACK IN THE WORLD

RAILWAY ACROSS THE NULLARBOR
The Trans-Australian railroad crosses the
Nullarbor Plain, named for its lack of trees.
This transcontinental Sydney–Perth line
opened in 1970, after the gauges in South
and Western Australia were standardized.

However, the attempt to develop a standard-gauge network would continue for much of the 20th century. The first concerted effort had been made in 1932, with the completion of a standard-gauge railroad between Sydney and Brisbane, but it was not until after World War II that real efforts were made to create a network across the whole country using the same gauge. In the 1950s, a plan was drawn up to create a national network of standard-gauge lines, but progress was slow because of the cost. It was not until 1962 that the Sydney–Melbourne route, the busiest long-distance line in the country, was standardized, and three years later the Sydney–Perth route followed suit. The Melbourne–Adelaide railway was converted to standard gauge in 1995, and the Adelaide–Darwin line followed nine years later.

There is no doubt that the gauge "bungle" cost Australia and its railroads dearly. Changing trains is both unpopular with passengers and costly for freight. The Australian experience is in great contrast to that other vast country developed thanks to immigration—the US, where the railroads were the principal catalyst for growth. But thanks to the Australian government's program of standardization and modernization—including the introduction of diesel trains and the closure of heavily loss-making branch lines—the railroads survived and, in places, even began to flourish, although they have never been as vital a part of the economy as in many other countries.

The experience of Australia is in sharp contrast to its neighboring country and fellow ex-British colony, New Zealand. Although the origin of the railroads was similar—the first lines were built by provincial governments, with the first one opening in 1863—central government took over in 1876. From then on, New Zealand's railroad system was developed in an integrated manner, using just one gauge. The government of the dominion of New Zealand deliberately expanded the railroads to provide a unifying force for the sparsely populated nation, which consists of two main

THE JOY OF TRAINS
The New Zealand government promoted rail travel not only for industrial purposes but for leisure activities too.

NEW ZEALAND RAILWAY REFRESHMENT ROOM
The refreshment rooms at provincial New Zealand Railway
stations often became the social heart of the townships
they served, offering hearty meals at decent prices.

islands. As the author of a history of New Zealand railroads, Neill
Atkinson, put it: "since the late nineteenth century, the state used its
expanding rail network to promote not just the development of
agriculture, industry, forestry, and mining, but to further policies in
areas as diverse as education, town planning, and recreation. Trains
ferried school children to the classroom, suburban workers to
factories and offices, sportspeople to competitions, and thousands of
picnickers and punters to beaches, parks, and racetracks."

Railroad stations soon became the hub of their communities. Each
station's "Railway Refreshment Rooms," sold appetizing, cheap food
and beverages served "in the legendarily thick New Zealand Railway"
cups, bowls, and plates. The railroads became an integral part of New
Zealand's economy and culture, and, by the early 1920s, more than 28
million journeys were made annually by rail in a country of a little
more than one million people. The railroads were, in the words of a
1938 advertisement for New Zealand Railways, "the industry that made
New Zealand—the people's railways for the people's profit." The
contrast with Australia could not have been greater.

High-speed
Steam Trains

THE CRUDE "TEA KETTLES ON WHEELS" built by pioneers, such as George Stephenson in Britain and Peter Cooper in the US were transformed over the first 100 years of the railroad as steam-train technology leapt forward. The period between the world wars, in particular, saw great engineers of the time turn steam locomotives into sophisticated powerhouses.

An interesting analysis undertaken in 1889 by early "train enthusiasts" E. Foxwell and T. C. Farrer provides a comparative analysis of the speeds of trains across the world. In general, the authors were greatly disappointed by the slowness of most trains. For their research, the pair included only "expresses," which they defined as averaging at least 29mph (46kph)—hardly speedy travel, but even so, not many qualifying trains were found. There were no such services in several major railroad nations—including India and the whole of South America—and only a few in Australia. Foxwell and Farrer found that the countries with the highest proportion of fast services were France and the Netherlands, which interestingly used mainly British locomotives. In Germany, an average of 35mph (56kph) was rare, while in Italy there was only one express—a daily service between Milan and Venice. Services in Sweden—then still a poverty-stricken agricultural country—were "poor," but Denmark had several good trains. Hungary, meanwhile, was praised for allowing the Orient Express (see pp.190–97) to average 32mph (51kph), which was faster than in neighboring Austria. In the US, the only trains that qualified as expresses were in the east, including the best service in the world—a train that ran the 40-mile (64-km) trip between Washington, DC, and Baltimore at an average of 53mph (85kph). Otherwise, Foxwell and Farrer were disappointed at the speed of many services in the US, finding that even those with famous names, which often contained words like "Flyer," barely qualified as express trains, averaging only around the 30-mph (48-kph) mark. The problem, they found, was that tracks in the US often had to go through the center of towns, where the trains had to travel very slowly because of potentially dangerous grade crossings.

LEADING THE WAY
Engineers board a steam locomotive on a French
railroad in 1870. The French led the race for greater
speed, calling their high-speed services "exprès"
trains (literally, "with purpose").

This intrepid pair of timetable-watchers returned to their task a
decade later and found that considerable improvements had been
made. By the start of the 20th century, France led the way with 20
daily expresses averaging at least 56mph (90kph) and a series of fast
international services running from Paris that covered much of
Western Europe, including Vienna and Warsaw. In Germany and
Britain, too, expresses routinely averaged 50mph (80kph) or more.

Part of the reason for these increases in speed was a trend that had
begun around the time of Foxwell and Farrer's first survey—rival
railroad companies would vie for the fastest journey time between
two points. The first such contest took place in Britain, in a bid for the
best time between London and Scotland. Two parallel lines, the East
Coast and West Coast main lines, traveled between these two
destinations. The companies that ran these services had a tacit
agreement that journey times between London and Edinburgh would
be 9 hours by the East Coast, and 10 by the longer West Coast route,
which also had to contend with steeper grades. But in June 1888, the
two West Coast companies—the London and North Western and
the Caledonian, which together provided a joint service—announced
that they would cut the extra hour from their service. A few weeks
later, just ahead of the Scottish grouse-shooting season—which
started on August 11 and represented a lucrative period for the

LOCH KATRINE
via West Coast Route

THE SCENIC ROUTE
A poster for the London-to-Scotland
West Coast line c.1910. Competition
with the East Coast line reduced
journey times by several hours.

London–Scotland operators—the East Coast companies (the Great Northern, the North Eastern, and the North British) retaliated. They cut half an hour from their timetable by limiting stopping times, reducing the journey time to 8 hours and 30 minutes.

The North Western (the main company on the West Coast route) responded quickly, vowing to slash the journey time to 8 hours—a full 20 percent reduction on its previous 10-hour schedule. The East Coast companies hit back in this dramatic game of poker, promising a journey time of 7 hours and 45 minutes. As the battle reached its height, people gathered at the point of departure to see off the rival trains, and their performance was reported in the manner of weekend football games. Tower operators along the routes took special care not to slow down these prestigious trains, and teams of workers ensured that the tracks were up to standard. Eventually, the East Coast companies peaked with a run of just 7 hours and 30 minutes later that summer, but the contest concluded soon after as the rivals came to agree on a standard journey time at 8 hours and 30 minutes.

Seven years later, in the summer of 1895—following the completion of the Forth Bridge, which greatly reduced journey times to the north of Scotland—an even fiercer and ultimately more dangerous race broke out. This time the trains ran by night, and the "race course" was extended to Aberdeen—over 500 miles (800km) from London by rail, and about 100 miles (160km) further than Edinburgh, the previous terminus. For added excitement, the trains from the two rival lines were forced to share the final section of track after the Kinnaber junction, 38 miles (61km) south of Aberdeen. The first train to reach this point was the clear winner. The contest lasted for 17 days in August

1895 and attracted huge crowds at the departure and arrival points. Rival companies engaged in various underhand methods, such as not stopping at intermediate points, traveling with just two or three cars to keep the weight down, and simply ignoring the timetable altogether. By the end, the times were incredible, with the East Coast service a mere 8 hours and 40 minutes. The West Coast companies finally managed to beat this by 8 minutes with an average speed of 63mph (101kph). However, the affluent passengers arriving for the season's grouse shooting did not welcome being turned out of their comfortable cars at Aberdeen at 5am, rather than 7am, which had been perfectly timed for breakfast.

Concerns over safety and cost caused the contest to peter out. Furthermore, the following summer a major disaster occurred at Preston on the West Coast line due to excessive speed. An inexperienced engineer failed to slow the train as it passed through the station, where a curve demanded a speed restriction of 10mph (16kph). Unlike most trains passing through Preston, the service was scheduled to run through without stopping, and the train jumped the tracks at 45mph (72kph). There was only one fatality, but the accident alerted passengers and the railroad companies to the risks of focusing purely on speed.

By 1900, the Great Western Railway (GWR), the biggest of the British companies in terms of mileage, was leading the race for speed. In 1904, one of its new locomotives, *City of Truro*, was clocked at 102mph (164kph) on a downhill stretch in Somerset. Although the exact speed is disputed, it is generally considered one of the first times in the world that the 100mph (160kph) barrier was breached. This occurred during a campaign by GWR to establish itself as the premier railroad company in Britain. The company's ambition led to competition with the London and South Western (LSWR) company over traffic from liners sailing to and from the US. Transatlantic ocean liners had traditionally docked at the English port of Southampton, with passengers then traveling onward to London by train. However, to save time, since travel by land

"… the traveling was so curiously smooth that [...] it was difficult to believe we were moving at all…"

CHARLES ROUS-MARTEN, ON THE *CITY OF TRURO*'S RECORD-BREAKING RUN

is so much faster than by sea, ships also began docking at the port of Plymouth—over 150 miles (240km) west of Southampton. Though further from London, passengers could rapidly cover the remaining distance by train, shaving nearly a day off the total journey. The LSWR had traditionally served Plymouth, but now the GWR also wanted a slice of the action. A full-scale war broke out as rival companies ran trains with no particular timetable, simply picking up passengers from ships as they arrived and steaming to London as quickly as possible. This was to have tragic consequences on June 30, when a special service that had left Plymouth just before midnight attempted to travel through the city of Salisbury at more than twice the 30mph (48kph) speed restriction and came off the rails. Of the 43 passengers on board, 24 were killed. While these special services were later run with more care, the race between the two companies continued until 1910.

In the US, the race for the fastest time between New York and Chicago lasted many years. It had begun in 1887, when the Pennsylvania Railroad introduced the *Pennsylvania Limited*—an all-Pullman affair boasting a barber's shop, valet, and maid service. Two years later the New York Central Railroad responded with a train that covered the 436 miles (701km) between New York and Buffalo in 7 hours, at an average speed of 61mph (98kph). Then in 1902, the Central hit back with the launch of the *Twentieth Century Limited* express train, which covered the near-1,000-mile (1,600-km) distance between New York and Chicago in 20 hours—a reduction of 4 hours on the usual time. In response, the "Pennsy," as it was known, renamed the *Pennsylvania Limited* the *Pennsylvania Special*, and managed to complete the journey in the same time as its Central rival. A battle ensued, with each company repeatedly trying to reduce the journey time in a series of much-publicized initiatives. However, these grand contests proved too costly for both companies, and eventually a gentlemen's agreement of a 20-hour journey time was reached.

Train races of this kind largely died down until the 1930s, when they were revived as a last-gasp attempt to help steam technology see off competition from rival methods of traction. Steam engines had improved remarkably between the wars thanks to several illustrious engineers. The greatest of these was Frenchman André Chapelon, whose rigorous scientific analysis and emphasis on efficiency were widely imitated, leading to radical improvements in locomotive performance. In Britain, a contest broke out in 1937 between the two consolidated companies—Britain's many companies had been merged into just four in 1923—serving the Scottish route. William Stanier, chief engineer at the London, Midland, and Scottish (LMS), built the Princess Coronation class of high-speed locomotives, which are widely recognized as the best British locomotives ever produced. On a specially arranged press trip, the first engine of the class reached 114mph (183kph), a speed that was intended to better the effort of its rival, the London and North Eastern Railway's (LNER) streamlined A4 Pacifics. The LNER claimed that an A4 had reached 113mph (181kph), but when the company learned the record had been beaten, it planned a record run in great secrecy. This was eventually undertaken in 1938 when *Mallard*, a streamlined A4, reached 126⅞mph (202.6kph) in a specially organized run on a straight piece of downhill track south of Grantham. This record for steam locomotives would never be surpassed, although a German Class 05 locomotive had come close two years earlier when it reached 124½mph (200.4kph) between Hamburg and Berlin. In the US, the use of high-speed diesels (see pp.224–31) on prestigious routes spelled the end of steam, and, by the 1970s, steam power in continental Europe had mostly been displaced by large-scale electrification.

RECORD-SETTER
In 1905, the steam locomotive PRR 7002 set a record time for the *Pennsylvania Special* service from New York to Chicago.

THE FASTEST-EVER STEAM TRAIN
The A4 Pacific-class *Mallard* charges through the English countryside. On January 3, 1938, it reached 126mph (203kph)—the fastest-ever speed for a steam locomotive.

Going Diesel: from the *Fliegende Hamburger* to the Future

STEAM ENGINES WERE DIRTY BEASTS—difficult to maintain, temperamental, and inefficient—so a search for alternatives started early. Experiments with electric traction led the way, but once the internal combustion engine had been invented, attempts to apply it on the railroads were bound to follow. The first internal combustion engines tried out on trains were fueled by gasoline, but it did not prove efficient and was expensive, especially for large engines.

German engineer Rudolf Diesel invented and patented the eponymous Diesel engine in 1892. Instead of a spark plug, it used air heated by high compression to ignite the fuel. Two main types of diesel engine are now in use: one provides power directly; the other, the diesel-electric, uses a diesel engine to power a generator, which then provides electricity for propulsion. Both have been used extensively.

Following Diesel's invention, experiments with locomotives began almost at once. However, there were numerous technical obstacles to overcome before they could be put into practical use, and, apart from a few appearing on a small Swedish railroad, diesel engines were not introduced until after World War I.

Efforts to build functional diesel locomotives continued through the 1920s. Despite some success on the US and Canadian railroads, the real pioneer was Germany, which began experimenting with powerful diesel engines for rail propulsion. The result was a two-car unit called the *Fliegende Hamburger* ("Flying Hamburger"), which represented a considerable advance for rail technology in both speed and efficiency. The Germans had, in fact, already established the world's rail speed record with a bizarre-looking four-wheel coach powered by a gasoline aircraft engine with an airplane-type propeller at the back. It was built by BMW and reached 143mph (230kph) on a test run in June 1931, but a host of technical difficulties ensured it never saw regular service.

By contrast, the *Fliegende Hamburger* did become a widely used model on several routes, first going into service under this name in the winter of 1932–33. It covered the 178 miles (286km) between Berlin and Hamburg in 2 hours and 20 minutes—an average speed of 76mph (122kph)—which required cruising speeds of around 100mph (160kph)

THE *FLIEGENDE HAMBURGER*
The streamlined design of the *Fliegende Hamburger* improved the train's speed and efficiency by reducing wind resistance, but also created an enduring design style.

to maintain the timetable, making it by far the fastest rail service in the world. The train had a remarkable streamlined design, like a Zeppelin airship, the result of wind-resistance tests in a wind tunnel. Although Hitler preferred automobiles, which he saw as the transportation of the future, the *Fliegende Hamburger* became part of his propaganda exercise to show the superiority of the "Thousand-Year Reich." The success of the *Fliegende Hamburger* soon led to the design being used on other services: two years later, a similar service was introduced on the Berlin–Cologne route, with an average speed of 82mph (132kph), and a *Fliegende Frankfurter* ("Flying Frankfurter") service followed. These diesel trains represented a radical technological development, but were laid up in World War II because of fuel shortages and only saw service again briefly after the war.

It was in the United States that diesel technology was developed more widely and successfully than elsewhere. Its development took place against the background of a need to compete in a nation where automobiles and later planes were eroding rail's market share. The US railroad companies had been taken over by the government during World War I because of their incompetence and refusal to cooperate with one another. They emerged from state control eager to improve what they offered, by using prestigious trains such as the *Pennsylvania Special* and *Twentieth Century Limited* as their trademark services.

By the late 1920s, however, these services had begun to seem slow, and their proprietors were desperately seeking a new technology to speed the trains up. This was particularly true for the railroad companies whose services crossed the vast swathes of the West. With car use still not widespread and aviation in its infancy (both the Model T Ford, the first affordable American automobile, and the first commercial domestic flight, between Boston and New York, were launched in 1927) and planes still posing a safety risk, the railroad companies began to look to diesel as the answer to their problems.

These new diesel trains were a different kind of train to the steam-hauled services. Consisting of perhaps half a dozen or eight cars, they were exclusively for passengers and provided a high degree of comfort. They were built of light stainless steel and alloys, which made them look sleek—especially compared to the heavy, conventional steam trains—and they ran fast between major cities, with limited stops to improve journey times.

While some diesel locomotives had already been introduced by various US railroads, these were confined to switching service since the powerful diesel engines were thought too heavy to be economical compared with traditional steam. Gasoline engines continued to be tried on some trains, such as the three-car *Blue Bird* trains of the Chicago Great Western Railroad that ran between the Twin Cities (St. Paul and Minneapolis) and Rochester, Minnesota—but again, they were simply too expensive to operate.

A key technical breakthrough for diesel engines was made by General Motors, the automobile company, which used alloys rather than steel to give a better power-to-weight ratio. The new, lighter engine attracted the attention of Ralph Budd, the head of the Chicago, Burlington, and Quincy Railroad—the Burlington, as it was known—at the Chicago Fair of 1932. Budd, one of the few railroad visionaries of the interwar period, realized the more powerful engine's potential to revolutionize long-distance rail travel. After seeing the engine at the fair, he commissioned a new type of streamlined diesel locomotive whose sleek and elegant looks were an attraction in themselves. Called the *Pioneer Zephyr* (the name *Zephyr*, meaning light west wind, came from Chaucer's *Canterbury Tales*, which Budd had been reading), it was launched amid much fanfare in May

BURLINGTON *PIONEER ZEPHYR*
The streamlined American *Zephyrs* gave the German trains a good run for their money in terms of speed and mimicked their gleaming, aerodynamic shapes.

1934 with a record-breaking "Dawn to Dusk Run" between Denver, Colorado, and Chicago. The 1,000-mile (1,600-km) trip was covered at an average of 78mph (126kph), remarkable by US standards and almost as good as the German diesels. However, a lot of special measures had been taken: patrol staff were placed on all 1,689 grade crossings along the route to stop car traffic well ahead of the train, and the train was limited to just three cars to reduce the weight. Famously, Budd reported that the fuel for the train was far cheaper than coal, costing a mere $14.64—although, at 4 cents per 1 gallon (3.8 liters), that meant 366 gallons (1,386 liters).

While such fast speeds could not be achieved on regular journeys, the train was far faster than any before and became the pioneer for a host of services that transformed American long-distance train travel in the years running up to World War II. A veritable family of *Zephyrs* and other trains sprang up on routes across the expanses of the West and along the East Coast, operated by these elegant new streamlined diesels. The Burlington, spurred on by Budd's enthusiasm, introduced in quick succession the *Twin Cities Zephyr,* between the twin cities and Chicago, and the *Mark Twain Zephyr* to St. Louis.

The Union Pacific's pioneer train went on a nationwide tour in 1934 before entering active service as the *Kansas Streamliner.* Renamed the *City of Salina* in 1936, it could run at more than 90mph (145kph) for long distances, averaging a stunning 92mph (148kph) on its run across the Nebraskan plains. The *pièce de résistance* of the period was the record-setting coast-to-coast journey by the *City of Portland,* a Union Pacific six-car sleeper train, which traveled the 3,250 miles (5,230km) between New York and Los Angeles in 56 hours and 55 minutes—nearly a day

506—Santa Fe's "Super Chief" Traveling thru the Orange Groves, California

TRAIN OF THE STARS
The Santa Fe *Super Chief,* which was launched in 1936 and ran between Los Angeles and Chicago, was known as the "train of the stars" because it was said to be Hollywood actors' favorite mode of transportation.

AVERAGE SPEED OF THE *FLIEGENDE HAMBURGER*	AVERAGE SPEED OF THE *CITY OF SALINA*
# 76 MPH (122 KPH)	# 92 MPH (148 KPH)

faster than the regular coast-to-coast service. Moreover, the fuel cost was only $80, compared with $280 for coal. This was only a trial run, however, and a nonstop coast-to-coast service was never established as passengers had to change trains at Chicago or St. Louis.

These modern diesels later acquired observation cars and other amenities to satisfy their affluent clientele. A fabulous variety of food and drink was on offer and for a while the trains became the envy of the world, with the companies competing to provide the best facilities. These trains were the height of America's rail system, indeed, they were probably the best the world had ever seen, but by the 1950s, as flying became safer and cheaper, they began to be phased out. Diesel remains, however, the main form of traction on US railroads to this day, with only a very small proportion of electrified railroads. Indeed, the typical image of American rail is of freight trains more than 100 cars long, being hauled by three or four powerful diesel locomotives.

Elsewhere in the world, diesels began to replace steam locomotives soon after World War II. Diesel multiple units, in particular, were a great way of saving money on branch lines. These had engines under the floor of the car, so that there was no need for a locomotive, and they could be driven from either end, obviating the need to turn around. The French even developed a rubber-tired diesel train called the *Micheline*, which was used extensively on minor routes.

However, electrification was often chosen over diesel. While it is initially more expensive, electric haulage is ultimately cheaper, cleaner, and offers faster acceleration. On many suburban services, steam trains were replaced directly with electric trains rather than diesels. Nevertheless, with only Switzerland operating a 100 percent electrified network, diesel remains an important form of traction, in particular for heavy freight and on little-used lines where the cost of electrification is not worthwhile. Diesel trains will be with us for a long time yet.

Diesel Power Meets Electricity

Diesel-electric traction emerged as the economically and functionally superior successor to steam power in the 1940s. Power is derived from a diesel engine (the "prime mover"), but is transformed into electricity by a generator that powers motors in the trucks, which in turn propel the train.

MARYLAND AND PENNSYLVANIA NO.81 (1946)
No.81, a General Motors EMD NW2-class switcher, was one of the first diesel-electrics to see widespread use. Popular due to its low cost and versatility, this small yet powerful class remains in service in small numbers even today.

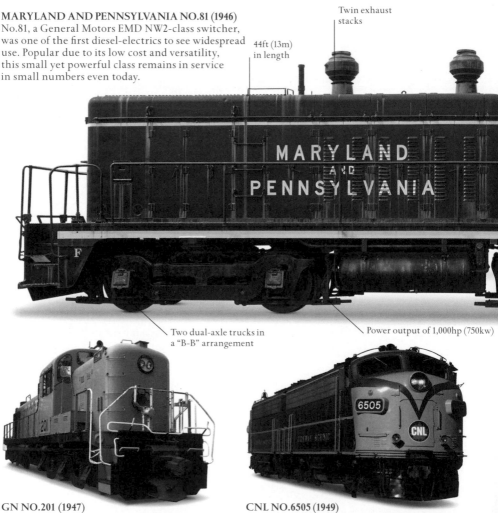

Twin exhaust stacks

44ft (13m) in length

Two dual-axle trucks in a "B-B" arrangement

Power output of 1,000hp (750kw)

GN NO.201 (1947)
This RS-2 class diesel-electric switcher was built by ALCO, and was one of 20 purchased by the Great Northern Railroad (US) to replace its coal-fired locomotives. Its 12-cylinder engine yields a power output of 1,500hp (1,100kw).

CNL NO.6505 (1949)
The EMD F7 diesel-electric was designed for freight, but was also used for passenger trains by some operators. Economical to operate and maintain, the pictured unit runs on the Conway Scenic Railroad (US) heritage line.

SP NO.6051 (1954)
One of nine EMD E9 diesel-electric passenger locomotives run by the Southern Pacific Railroad, No.6051 hauled services out of Los Angeles. Capable of a 2,400-hp (1,790-kW) power output, it was retired in 1969 and restored to its original "daylight" red-orange paint scheme.

ERIE LACKAWANNA NO.3607 (1967)
No.3607 was a General Motors EMD SD45, a six-axle diesel-electric freight locomotive with a power output of 3,600hp (2,680kw). More than a thousand units were produced between 1965 and 1971, a small number of which are still in operation on US railroads.

14ft 6in (4.4m) in height

Total weight of 124 tons (112 tonnes)

BRITISH RAILWAYS D200 (1958)
An English Electric Type 4 Class 40, D200 was one of the first wave of diesel-electrics to ply the rails in Britain. With a top speed of 90mph (140kph), the class had been intended for passenger expresses, but was relegated to slower passenger and freight services.

NS NO.9628 (1996)
Norfolk Southern Railroad (US) still operates a fleet of 1,090 General Electric Dash 9-40CW diesel-electric locomotives, the first of which was introduced in 1996. The class is powered by a 16-cylinder engine that is limited to a power output of 4,000hp (3,000kw) to improve running costs.

World War II:
Atrocities on the Line

ALTHOUGH THE RAILROADS' ROLE IN WORLD WAR II was not quite as fundamental as it had been during the World War I (see pp.270–75), they were nevertheless a vital part of the logistics of war. With gasoline in short supply and a lack of modernized roads in many conflict areas, the railroads carried troops and supplies far and wide. However, the railroads were also involved in two of the greatest war crimes of the World War II: the Holocaust, which resulted in the deaths of two thirds of all the Jews in Europe at the hands of the Third Reich, and the construction of the Burma (now Myanmar) to Siam (now Thailand) Railway, for which the Japanese used prisoners of war. These two events portray a darker side that is often left untold in railroad histories, but vividly, if hauntingly, demonstrates the power of the railroads and their significance to those who controlled them.

Approximately six million Jews were killed by the Nazis in the Holocaust, along with millions more from other groups such as Slavs, Poles, Romanies, communists, and homosexuals. Most of the victims were taken to death camps by train, and the sheer volume and speed of the deportations would not have been possible without intensive use of the railroads. Any other method of transportation would have presented insuperable problems—to have devoted so many trucks to the task would have damaged the Germans' war effort, and marching victims along roads might have revealed the true horrors of the Nazi's plans to the wider population.

The first of these trains were used principally to move German Jews to ghettos and ran between Germany and Poland (and then further east to Riga in Latvia). The grim dispatch of Jews and other sections of the population to the concentration and death camps began in the spring of 1942, and the flow intensified over the following two years. These deportations were carried out in a systematic manner on an industrial scale, as part of the "Final Solution" agreed at the notorious Wannsee conference in January 1942. Efficiency was seen as essential and required the active involvement of numerous German government ministries, including the Reich Security Main Office (RSHA), the Transport Ministry, and the Foreign Office, as well as the corresponding organizations in other allied or occupied states who were required to hand over their citizens. Many railroad workers were involved, too.

The deportations were mostly carried out in freight cars. Some victims, notably those from the Netherlands and Belgium, were transported in third class passenger cars, partly to maintain the subterfuge that they were merely being re-homed. Conditions on the trains were appalling: the freight cars were supposed to be filled with up to 50 people, but in fact, due to a shortage of cars, they sometimes had as many as 150 occupants, which meant standing room only. Trains could carry a maximum of only 55 cars each, as anything longer would travel too slowly. There was no food or water on the journey and only a bucket latrine. The only ventilation was through a barred window and consequently many people suffocated. In the summer, the temperature could be unbearably hot and victims baked, while in the winter the temperature plummeted and victims froze. The trains were given the lowest priority on the railroad network and consequently journeys were

AUSCHWITZ II—BIRKENAU
Trains carrying deportees entered this extermination camp in Poland on a specially built branch line that extended right up to the gates of the camp.

often delayed while more important military convoys were allowed through. This meant that the deportees were sometimes held in sidings for days and the average time for journeys, which should have taken a day or so, was four and a half days. The longest journey involved the deportation of Jews from the Greek island of Corfu, who were taken by boat to the Greek mainland and then transferred to a train. The train was held up several times and consequently took 18 days to reach Auschwitz. By then, many of the occupants were already dead. Given the conditions, length of the journey, and the lack of food and water, deaths in transit were common and most trains arrived containing several corpses.

One of the least known but cruelest parts of the deportation process was the fact that the victims were forced to buy tickets for the journey, a full one-way fare for adults, with children being charged half price. This scheme generated an astonishing amount of revenue, calculated at around 240 million Reichmarks (around $201 million). At the peak of the process, there were up to 10 trains per week arriving at the camps. For the Nazis, it was the very efficiency of this mode of transportation that made the extermination of so many people possible. The despatch of trains only began to slow down when the Allies invaded France in the summer of 1944 and the operation ceased entirely as the Third Reich began to fall apart in the spring of 1945. In the 21st century, several railroad companies have apologized for their role in these wartime deportations, including Dutch railways Nederlandse Spoorwegen in 2005 and French SNCF in 2011.

During the Holocaust the railroads conveyed millions to their deadly destinations. Meanwhile, in Asia it was the railroad line itself that was the scene of another war crime. When the Japanese overran Singapore, the main British naval base in Southeast Asia, in February 1942, they captured more than 80,000 British, Indian, and Australian troops. Along with 50,000 existing prisoners, many were sent to work on the construction of the Burma to Siam Railway, which was intended to provide the Japanese with a vital supply line as they advanced westward, toward India.

As there was no adequate existing road or rail links between the two countries and the sea route was vulnerable to attack from Allied ships and submarines, a railroad seemed the obvious solution. To

BRIDGE BUILDING
Prisoners build at bridge Tamarkan, 34 miles (55km) north of
Nong Pladuk in Siam. The scaffolding was made from bamboo,
but the bridge, completed in April 1943, was made of steel.

build the 300-mile (483-km) line, which went through harsh
mountainous territory and tropical rainforest, the Japanese used
forced labor of up to 330,000 men, mostly made up of conscripted
locals, but also including more than 60,000 Allied Prisoners of War.
The line was started simultaneously from both ends, Thanbyuzayat
in Burma and Nong Pladuk in Siam, in June 1942. There was a constant
shortage of materials and most of the equipment, including tracks
and ties, was brought from dismantled branches of other local
railroads. The human cost was, however, appalling and the line
became known as the "Death Railway." Dutch merchant sailor Fred
Seiker described his experience as PoW on the railway labor force:

> You carried a basket from the digging area to the top of the embankment,
> emptied it and down again to be filled for your next trip up the hill. Or
> you carried a stretcher—two bamboo poles pushed through an empty
> rice sack—one chap at each end, and off you went. Simple really. But in
> reality this job was far from easy. The slopes of the embankments
> consisted of loose earth, clambering to the top was a case of sliding and
> slithering with a weight of earth in attendance. This proved to be very
> tiring on thigh muscles and painful, often resulting in crippling cramp.
> You just had to stop, you could not move. Whenever this occurred the
> Japs were on you with their heavy sticks, and beat the living daylight out
> of you. Somehow you got going again, if only to escape the blows.

Working hours were typically 7:30am to 10pm and the food rations were just 7oz (200g) of rice per day, often with no vegetables, let alone meat. Robert Hardie, a British doctor who was captured in Singapore, described in diaries published posthumously in the 1980s how the Japanese would line up the sick that he was tending and demand that a dozen of them should be sent to the work camp. He wrote: "One is under constant pressure to provide men to work under this Nipponese system: for certain groups of men are given certain work to do in a certain time. If many go sick in a group, the others have to work all the harder and longer". Hardie also recounted the repeated refusals of his captors to provide even basic medical supplies, as well as their indifference to the spread of diseases, such as cholera, malaria, and dysentery. The death rate was particularly high in the final months of the railroad's construction as the Japanese were desperate for the line to be completed.

Despite the weakness of the men due to starvation and disease, their sheer numbers and the pressure from their Japanese guards ensured that the line was completed remarkably quickly, in just 16 months. On October 17, 1943, the two sections of the line met,

HELLFIRE PASS
The most difficult section of the line involved cutting through rock in a remote area of Siam. The work was extremely arduous, especially given the lack of adequate tools, and became known as Hellfire Pass.

"It didn't matter how many men died, the railway would be built. And it was built at a terrific cost"

JOHN LESLIE GRAHAM, BRITISH POW

11 miles (17km) south of the Three Pagodas Pass at Konkuita in Siam. By this time the death toll was estimated to have reached more than 100,000, a figure that included a quarter of all the Allied prisoners. The railroad immediately became a vital part of the Japanese line of communication after their navy lost control of the South China Sea during the summer of 1942. However, as an essential bridge to the Burmese railroad system was never completed, the supply route still entailed transporting some goods via ferry.

The story of the Burma to Siam Railway reached a wider audience through the 1957 David Lean film *The Bridge on the River Kwai*, which was based on a book by Pierre Boulle. The story refers to Bridge 277, built over a stretch of river that was then called Mae Klong. The tale is largely fiction, since it shows the bridge ultimately being destroyed by sabotage. In reality, the bridge remained in working condition, despite the men's attempt to undermine it by mixing the concrete poorly and encouraging termites to use wooden supports as nests. Although the film was criticized as unrealistic and failing to depict accurately the appalling conditions which the men lived and died under, it nevertheless helped to ensure that the memory of this terrible project lasted longer than the line itself. Much of the railroad was damaged during the war and it never fully reopened. A few sections, such as the line between Kanchanabur and Nong Pladuk, did reopen in the 1950s, but most of it was abandoned, submerged under water by the Vajiralongkorn dam, or reclaimed by the jungle. Today a daily tourist train runs along the surviving 130-mile (209-km) section, while other parts have been converted to a walking trail. Three large cemeteries honoring those who lost their lives building the line can be found along the route, along with an Australian-built memorial and museum at Hellfire Pass, and several smaller memorials.

TRACKS OF THE PAST
Tourists can still get a glimpse of some
of the unforgiving and perilous terrain
through which the Burma to Siam
Railway was constructed.

The Iron Road Today

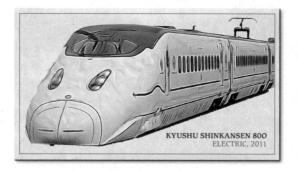

KYUSHU SHINKANSEN 800
ELECTRIC, 2011

In the immediate postwar years the railroads suffered a period of decline. The car had become the preeminent form of passenger transportation, and the semitruck dominated the freight business. Highways were being built across the world, and they were intended to replace the railroads. The plane had also become affordable, and both international and domestic services were proliferating globally. Railroad closures were inevitable, starting with branch lines and rural tracks, then spreading to the main lines themselves. In the United States, the decline of passenger railroads was particularly swift, and for a while it seemed as if they would barely last into the 21st century.

Then everything changed. The oil crisis of the mid-1970s, together with congestion on the roads and environmental concerns regarding exhaust emissions, suddenly brought the train back into fashion. In Japan, high-speed "bullet" trains were introduced, which traveled faster than any car, and networks of similar lines were built in France and Germany, and later in China and Spain. Moreover, when the long-proposed 31-mile (50-km) Channel Tunnel linking France and Britain was finally given the go-ahead, it was built to carry trains because a railroad was the only practical way of traveling in such a long tunnel. Similarly, to relieve the pressure of congestion on Europe's Alpine roads, a series of lengthy modern rail tunnels are now being built under the mountains. In 2006, China completed the highest railroad in the world—the Qinghai–Tibet line, which offers a far easier journey than the highly dangerous road—and is currently building the largest high-speed network in the world.

The railroads are therefore booming in the 21st century. Light rail systems, subways, high-speed lines, and newly electrified tracks are being built across the world. The 19th-century invention has found new friends in the 21st century and is enjoying a fantastic renaissance thanks to its ability to transport masses of people quickly and cheaply.

Brezhnev's Folly

O F ALL THE MANY HAREBRAINED IDEAS in the history of the railroads, one takes the prize for being not only the craziest but also the most costly. That winner is the 2,300-mile (3,700-km) Baikal-Amur Mainline, a branch of the Trans-Siberian Railway (see pp.180–89) that dwarfed the original railroad both in difficulty and cost. It was one of the most ambitious of numerous megaprojects dreamed up by the Soviet regime to demonstrate the superiority of Communism (others included the space program and a plan to reverse the flow of several Siberian rivers, which was fortunately abandoned). The BAM, as the railroad became universally known, took three quarters of a century to complete and cost $14 billion, but its worth still remains in doubt.

The idea behind the BAM was to provide an alternative route to the existing Trans-Siberian, which ran from European Russia to the Asian Pacific. The Soviet regime first mooted the railroad as a strategic alternative in the 1930s Stalinist era, following disputes with China and Japan over a section of the Trans-Siberian that crossed Chinese territory to reach Vladivostok. Even the Amur Railway, a longer route over exclusively Russian territory that was completed in 1916, was considered too close to the Chinese frontier and therefore vulnerable to attack. To counteract the perceived threat, Stalin's government passed a secret decree to construct a new line running parallel to the existing Trans-Siberian but around 500 miles (800km) farther north. No details of the route were set out other than the terminal at Tayshet, where the line diverged from the Trans-Siberian, and Sovetskaya Gavan, on the Pacific Ocean north of Vladivostok.

The land crossed by the proposed new railroad was virtually uninhabited, as most of the population of Siberia had settled within 100 miles (160km) of the Trans-Siberian, so the reasons for undertaking this vast project were (the Sino-Japanese menace notwithstanding) questionable. The Soviet government had started publishing five-year plans setting out its economic targets, and its Second Five-Year Plan, for 1933–37, emphasized the economic advantages of building the BAM. It "will traverse little-investigated regions of eastern Siberia and bring to life an enormous new territory and its colossal riches— amber, gold, coal—and also make possible the cultivation of great tracts of land suitable for agriculture," the plan stated.

BAIKAL-AMUR MAINLINE

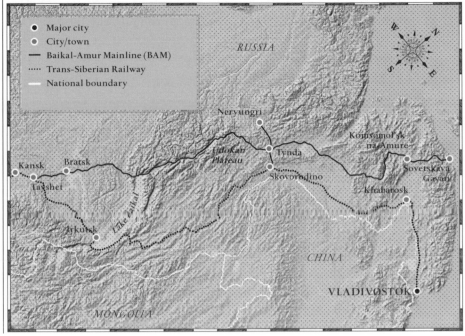

- ● Major city
- ○ City/town
- — Baikal-Amur Mainline (BAM)
- ⋯⋯ Trans-Siberian Railway
- ▬ National boundary

RUSSIA

Neryungri

Komsomol'sk-na-Amure

Udokan Plateau

Tynda

Bratsk

Kansk

Tayshet

Skovorodino

Sovetskaya Gavan'

Khabarovsk

Irkutsk

Lake Baikal

CHINA

VLADIVOSTOK

MONGOLIA

Surveyors were dispatched to the Siberian wastes to prospect the land. Working under a reign of terror, several were executed for not doing their job properly in the view of the regime, while others ended up in the construction gangs. These were made up of prisoners—mainly political—sent to the Gulag (the government's corrective labor camps) by Stalin's increasingly repressive regime. Tayshet became infamous as a camp for Gulag prisoners working on the railroad after Alexandre Solzhenitsyn described it in *The Gulag Archipelago*. Published first in the West in 1973, the book drew the world's attention to the atrocities that took place in Soviet forced labor camps under Stalin. Hundreds of thousands of prisoners were sent to camps in Siberia and forced to work on the line in appalling conditions—far worse than those endured on the Trans-Siberian. Russians knew that being sent to the "Bamlag"— the camps of the BAM Gulag—was a death sentence, as prisoners were subjected to unendurably harsh conditions and systematically starved.

The northern latitude of the track meant that most of the terrain it traversed was permafrost—a leftover of the freezing temperatures of the Ice Age—which presented particular difficulties for track-laying. Once railroad workers started digging through the insulating surface layer of soil to the permafrost, ice that had been frozen for millennia thawed and did not refreeze, even in winter. Instead, the land turned marshy and unstable. Waiting for the ground to settle would have delayed work by years, so tracks were laid regardless, resulting in rail breaks and derailments. Disturbing the permafrost also led to an increase in seismic activity in the region.

Not surprisingly, little progress was made on the line. On top of permafrost and earthquakes, the mountainous region north of Lake Baikal posed insurmountable difficulties. Weather conditions were extreme, with only 90 frost-free days a year, and winter temperatures as low as –76 °F (–60 °C), meant that mechanical equipment did not work and special cold-resistant steel had to be used for the rails. These factors, combined with a starving labor force, meant that by the outbreak of World War II, when work was halted, only a couple of short sections at each end had been completed.

Remarkably, as soon as the war ended in 1945, work resumed on this vain project, which Stalin seemed intent on completing. The workforce was now made up of Japanese and German prisoners of war, who were treated more harshly even than the prewar domestic convicts. It is

CONDITIONS IN THE GULAG
This ink drawing by a Gulag inmate gives a sense of
the conditions endured by forced laborers working on
the BAM—cold, hungry, and worked to the bone.

"BAM will be constructed with clean hands"

LEONID BREZHNEV

estimated that of 100,000 German POWs sent to the Ozerlag camp near Lake Baikal, only 10 percent survived to be repatriated in 1955; the Japanese prisoners suffered similar rates of mortality. A conservative estimate of the death toll of the two groups is 150,000—and all for nothing, since barely 450 miles (725km) of the line had been completed by the time work was again halted following Stalin's death in 1953.

The demise of Stalin and his repressions might have signaled the end of the BAM project. Nikita Khrushchev, Stalin's successor, showed no interest in the railroad, and the Gulag camps that had provided its labor were closed. By the 1960s, however, interest was revived, with a series of ostensible new reasons to build it: the line would relieve congestion on the Trans-Siberian, open up gas fields in western Siberia, and provide a new route for burgeoning container traffic between the Far East and Europe. Moreover, vast copper deposits had been discovered at Udokan, 250 miles (400km) east of Lake Baikal.

It was Leonid Brezhnev, the uninspiring and deeply conservative leader whose stern demeanor characterized Soviet rule in the 1970s, who decided to restart the program. Now a new type of cheap labor was to be used: volunteers of the "All-Union Leninist Youth League" (the Komsomol). The project was turned into a propaganda exercise, in the hands of the Komsomol, not only to demonstrate the advantages of the Communist way, but also to enthuse a whole generation of young people with that ideal. By helping to build the line, the volunteers would become lifelong supporters of the regime. Building the BAM became a rallying cry for Socialist propaganda as well as a path to victory over the obstacles that nature and the elements posed against humankind, making it a struggle that had to be won, at whatever cost—and that cost was to prove enormous.

In 1972, after much secret preparation, the Komsomol announced that work on the railroad would restart immediately and the BAM would be completed in 10 years (later extended by two years to 1984). The project was given priority over other Soviet plans and a nationwide appeal for volunteers was issued. While some young people may have turned up at the recruitment offices for idealistic reasons, there were

WORKERS' PRIZE
Idealistic young Komsomol workers on
the BAM were rewarded with medals
and praise, as well as material goods
such as cars and housing.

practical incentives too: volunteers were promised
priority allocation of housing and cars, both of which
were in short supply in Soviet Russia, and for those interested
in a political career, working on the BAM was essential for
their résumés. The pay was good too, compared with
jobs back home. However, even those who started off
with genuine enthusiasm were soon disillusioned. They
were well treated, but once they discovered the sheer
scale of the project and the incompetent way in which it
was being run, it dawned on them that the railroad was no advertisement
for the Communist system—quite the opposite. They realized, as
geologists and other scientists had long known, that creating a railroad
in northern Siberia was not a good idea.

There were innumerable practical difficulties. The recruits were
given little training for what was a skilled task, no detailed route had
been prepared, and the physical conditions were even worse than
expected. Attempts to continue work in winter, because of the
ideological need for rapid progress, were counterproductive. Although
work was stopped when temperatures reached −4 °F (−20 °C), even
above that temperature bulldozers stopped functioning and axes
shattered. Any lessons learned from previous attempts to work in the
permafrost seemed to have been forgotten, and once again whole
sections of track gradually sank into the morass, while station and
warehouse buildings constructed on shaky foundations collapsed.

Conditions on completed sections of the line were so bad that trains
had to go extremely slowly and derailments were frequent. One
117-mile (188-km) section between Tayshet and Tynda that opened early
took eight hours to traverse. With the tunnels also proving far more
difficult to dig out than expected, the opening of the line was inevitably
delayed: the 10-year target had always been a fantasy. The 9-mile
(15-km) Severomuysky Tunnel, east of the lake, caused almost
insuperable problems. When digging began in 1977, water from an
underground lake flooded it. Although an ingenious solution was
eventually found—liquid nitrogen was injected into the tunnel walls,

GLORY TO THE CONSTRUCTORS OF BAM RAILWAY!
Brezhnev's decision to finish building the BAM in the 1970s
was a monumental propaganda exercise, intended to inspire
a new generation of workers with the ideals of Communism
and gain their support for the regime.

freezing the water while the tunnel was lined with a concrete shell—it took 26 years to complete. In the meantime, two very steep bypasses were built, both adding considerably to the eventual journey time. Nevertheless, the Communist authorities—intent on using the line for propaganda purposes—stuck to the 1984 opening date with a ceremony that featured the hammering fast of a golden spike to set the final rail. It was a complete sham. No foreign journalists were invited, as it would have been obvious that the line was far from complete.

In the end, the BAM was officially opened three times. Brezhnev had died in 1982, but Mikhail Gorbachev, who succeeded him in 1985, continued with the program. It was by then soaking up one percent of the nation's total annual GDP. Seven years after the first ceremony, Gorbachev announced that the line was open and boasted that it would form a new link between Russia and Japan. Even then, the intractable Severomuysky Tunnel was still not complete and several sections could only accommodate slow trains used to supply materials for constructing the line. Russia's post-Soviet president Vladimir Putin announced the line's completion in 2001, and the tunnel finally opened in 2003.

The BAM has not lived up to Brezhnev's expectations. The promise of opening up a vast agricultural region was always a delusion, since the Siberian climate is so harsh. It did not relieve pressure on the Trans-Siberian, since it is the section west of the junction at Tayshet, which is shared with the BAM, that is under greatest pressure. Nor has it provided a practical alternative route between Asia and Europe.

Just as the Trans-Siberian contributed to the downfall of the Czarist regime, the BAM and the excessive resources devoted to it helped to bring down Communism. Most of the half a million Komsomol volunteers and other workers who built the line returned to their home towns deeply sceptical of the ideals of Communism and its grand projects. Indeed, many were infuriated that the cars and housing they had been promised were denied to them, a failure that led to demonstrations in the post-Soviet era, with ex-BAM workers demanding that their vouchers for volunteering be redeemed.

Far from carrying people to the promised land of a 21st-century future, as the slogans had promised, the BAM clearly went nowhere. The BAM became, in the Soviet era, the butt of popular jokes, symbolizing failure and the powerlessness of the Soviet leaders. As Christopher J. Ward wrote in *Brezhnev's Folly*, a history of the line:

A FREIGHT TRAIN THUNDERS THROUGH SIBERIA
BAM trains pass through the virtually uninhabited snowy
wilderness of northern Siberia. Passengers are few, but
container traffic is gradually increasing.

By repeating ad nauseam claims of BAM's economic, social, and cultural
significance, the Komsomol, the Communist Party, and the Soviet
government held an unwavering belief that the USSR's youth needed
this message to avoid a loss of collective faith. Ironically, however, the
realities of the railway helped to intensify such a loss of faith in the Soviet
political and economic system in general.

Today, however, there are a few signs that the railroad may have been
worth at least some of the effort. Russian Railroads, the state-owned
company, plans to increase container traffic on the line, which is now
carrying more minerals from Siberian mines too, and there has been
further recent investment, including the construction of the
Kuznetsovsky Tunnel, at a cost of $9million for less than 2½ miles
(4km), to give better access to the Pacific, as well as improvement of
parts of the line that were slow due to steep grades. However, the BAM
will still go down in history as one of the craziest civil engineering
projects ever attempted and forever be known as "Brezhnev's folly".

GIANT FREIGHT TRAINS
Freight trains are the most energy-efficient form of transportation for moving large quantities of goods over long distances. The biggest ever recorded—which took iron ore from a mine in Western Australia in 2001—was over 4½ miles (7.3km) long.

Railroads Lost and Found

A LMOST AS SOON AS THE RAILROADS WERE BUILT, sections started to become redundant and were closed. Some lines, built by railroad companies to meet a need that turned out not to exist, were never profitable and were soon abandoned; others, built to serve a particular mine or factory, lost their purpose when the works in question closed. The rapid demise of some superfluous early railroads was inevitable. In Britain, 200 miles (320km) of track had already closed by World War I, either because the lines were never viable or because of competition from rival routes. The first significant closure was the 17-mile (27-km) line between Chesterford and Six Mile Bottom in Cambridgeshire, built by the Newmarket and Chesterford Railway. It was closed in 1851 after a more direct route was created. Unusually, the line closed outright rather than being kept open as a freight route, a profitable alternative for many early lines. Other countries simply built too many railroads. France had more railroads than it needed, thanks to government support for railroad construction in the 1860s, and many of the lines did not survive into the 20th century. Ireland, despite enduring the terrible famine of the 1840s that resulted in the death and emigration of 3 million people, built a network of lines that never made economic sense, given the country's sparse population. At the 1920s peak, it boasted 3,440 miles (5,540km) of railroads serving a population of 4 million. It was hardly surprising much of this network closed in the 1930s, especially as automobile use increased.

In a way, it was remarkable that more lines did not close in this period, one reason being that railroad companies had to seek government permission to close lines in many countries, including the US and Britain. Permission was rarely granted since terminating a service was inevitably unpopular and attracted negative publicity for politicians, meaning that many little-used lines remained open. In Britain, an 1844 Act required railroads to run at least one service a day to serve poorer travelers. This resulted in a phenomenon known as "parliamentary trains," with companies operating the minimum number of trains necessary to circumvent the costs of closure. Some of these "ghost trains" run to this day. Even rather hopeless lines could be kept open to avoid controversy. Renowned chronicler of the

Victorian railways Jack Simmons wrote of the Bishop's Castle Railway in mid-Wales, for example, which became insolvent five years after its completion in 1861 and yet remained in operation until 1935.

Relatively few railroads closed around the world in the interwar period. In part this was because of local opposition to closure, as the railroads were now an established form of public transportation—usually the only one until the advent of buses. Railroad companies were reluctant to sanction closures too, as it was hard for them to predict the effect of closing branch lines on the network as a whole. Railroad economics are complex, and managers had to determine whether passengers on a branch line would continue to use the system if their local service was no longer provided before deciding if the revenue they contributed overall made it worth keeping an unprofitable branch line open. This is a continuing conundrum.

In Britain, which boasted over 20,000 miles (32,000km) of railroads at their peak before World War I, there were few closures in this period, despite growing competition from buses and automobiles. Just 240 miles (386km) of track were shut and another 1,000 miles (1,600km) restricted to freight services. Only the most extreme cases succumbed: the Invergarry and Fort Augustus branch line in the Scottish Highlands, which was completed in 1903 but never reached Inverness as intended, was closed in 1933 because it was carrying only six people per day. Closures remained the exception, however, until after World War II.

In the US, one type of train suffered major closures in the interwar years. This was the interurban (see pp.56–57), an extended railroad system that linked neighboring towns on tracks alongside roads. Built cheaply, they sprang up across the nation in the 1910s, usually operating a single car, akin to a bus on tracks. At their peak in 1913, these crude railroads covered 15,000 miles (24,000km), but they were badly hit by competition from automobiles. Closures started immediately after World War I and accelerated in the Depression after 1929, meaning most interurban lines had a lifespan of barely 20–30 years—very short for the cost of the infrastructure. A remarkable 6,350 miles (10,250km) were abandoned in the early 1930s, and by the end of World War II

THE LENGTH OF US RAIL TRACK CLOSED BETWEEN 1916 AND THE PRESENT DAY

160,000 MILES
(258,000 KM)

only a few lines survived. All but a handful had shut by 1960. As George Hilton and John Due, authors of *The Electric Interurban Railways in America*, put it: "the interurbans... never enjoyed a prolonged period of prosperity... they played out their life cycle in a shorter period than any other important American industry."

In almost every country with a significant railroad system, there were line closures in the postwar period, and the US was no exception, closing lines in great numbers. Major railroad companies realized their profits would now come from freight rather than passengers, given the growing competition for suburban journeys from automobiles and for long-distance travel from planes. Closures required permission from the Interstate Commerce Commission, the federal agency that regulated the railroads. By the early 1960s, all the main companies were petitioning the commission to close lines, and by the end of the decade there was a veritable stampede to end passenger services. Many companies used subterfuge to force the commission's hand, using old rolling stock, reducing services, and even demolishing stations to make lines appear uneconomic. Once permission to close was forthcoming, the companies were ruthless about implementing closure. So eager were they to get out of the passenger business, they stopped trains from running the minute authorization came through—so commuters of the Chicago Aurora and Elgin Railroad who had taken the train into town on the day closure was confirmed had to make their own way home. The Louisville and Nashville Railroad dumped its last 14 passengers in Birmingham, Alabama, 400 miles (645km) short of their destination; only after protests was a bus provided to get them to their final destination.

The pace of US closures became so frantic that the government intervened, creating the state-owned and subsidized railroad company Amtrak in 1971 to safeguard the remaining passenger services. It is an irony that the US, which tends to eschew state intervention, still has a nationalized railroad system. Although it went from a peak of 254,000 miles (410,000km) in 1916 to 94,000 miles (152,000km) today, it is still the biggest railroad system in the world, although it is mainly used for freight. Amtrak carries a mere 30 million passengers annually, compared with roughly 1.2 billion in Britain and 1.1 billion in France, both countries with far smaller populations.

In Britain, pressure to close lines came to a head in 1963 with the setting out of a radical plan by the chairman of the British Railways Board, Richard Beeching. He found that the railroad network was very unbalanced, with half of the country's 18,000 miles (29,000km) of track carrying only 4 percent of passengers. His solution was radical. The now infamous Beeching Report, known by its detractors as Beeching's Axe, resulted in the closure of 5,000 miles (8,000km) of track and a third of the 7,000 stations. While many lines that were clearly not viable were axed, some major routes were terminated too— subsequently a great cause of regret.

In other countries, closures were carried out incrementally, less radically than in the US and more slowly than in Britain. France lost half of its 37,000-mile (59,000-km) network in closures that started in the 1930s until, to protect the remaining lines, the government nationalized the system in 1938. In Eastern Europe, Communist regimes retained their rail systems more-or-less intact as few people could afford cars, but extensive closures took place after the Iron Curtain fell in 1989. However, in recent years the closure process has been reversed in some countries and railroad mileage has increased, with the reopening of mothballed lines and the construction of new, principally high-speed lines.

THE BEECHING REPORT
Richard Beeching, author of the report that drastically reduced Britain's railroad network, became a hate figure for railroad advocates.

All these closures had one surprising, beneficial side effect—the creation of a preserved railroads industry based on steam trains. Countries where steam trains had until recently remained in service, including Poland, China, and India, attracted enthusiasts; in other countries, volunteers preserved sections of old lines and now operate tourist services on them. The first railroad in the world to be preserved as a heritage railroad was the British Talyllyn Railway, a 7-mile (4-km)

"Beeching's report marked the end of our romance with the train, and the rise of the car"

IAN HISLOP, EDITOR OF *PRIVATE EYE* AND RAILROAD ENTHUSIAST

narrow-gauge railroad opened in 1866 to carry slate down a Welsh valley from quarries to the coast. When its closure was announced in 1951, a group of volunteers kept the line open. The railroad heritage movement has spread around the world since then, and some of these lines even operate regular services that connect with the main line. Old lines are being reopened all the time, among them the Welsh Highland Railway, which reopened in 2011 and now, together with the Ffestiniog Railway, provides an 50-mile (80-km) route across north Wales. In France, too, there are around a hundred such preserved lines, covering 750 miles (1,200km) of track. These railroads attract 3 million visitors a year, and include a section of 2-ft (60-cm) narrow-gauge lines used on the Somme front in World War I (see pp.278–87). The US also has a big preservation movement. Its most scenic railroad is the Durango and Silverton Narrow Gauge Railroad, a 45-mile (73-km) line through Colorado. The route was built in 1881–82 by the Denver and Rio Grande Railroad to exploit silver and gold mines in the San Juan Mountains. The remaining section is one of the few in the US that has seen continuous use of steam locomotives since the 19th century.

Some of the most spectacular railroads in the world have been brought back into use after being abandoned. In Ecuador, a 280-mile (450-km) line between the port city of Guayaquil and the highland capital, Quito, has been restored and its steam locomotives are now a major tourist attraction, recreating one of the greatest rail journeys in the world. Thanks to these preserved railroads and the volunteers who work on them, the steam engine will remain a source of admiration for generations to come.

DURANGO AND SILVERTON RAILROAD
The Durango and Silverton Narrow Gauge Railroad owes its preservation in part to its Hollywood success: it has featured in numerous movies, among them *Viva Zapata* and *Butch Cassidy and the Sundance Kid*.

WHITE PASS AND YUKON RAILROAD
Built in 1898 during the Klondike Gold
Rush, the WP&YR suspended operations
in 1982, when the mines closed, but was
resurrected as a tourist route from Skagway,
Alaska, to Carcross, Yukon, Canada in 1988.

Vive le Channel Tunnel

THE OPENING OF A TUNNEL LINKING Britain and France in 1994 was the culmination of more than 180 years of debates, discussions, and delays. The first proposal for a tunnel under the English Channel (although not a rail one) had been made as far back as 1802, by one of Napoleon's engineers. At that time, the journey from London to Paris took about four days (or several weeks if the Channel winds were unfavorable) but the British military, politicians, and even the press were not sold on the idea of a tunnel, fearing that it would leave the island nation vulnerable to attack. Over the next half century, various plans were proposed, on both sides of the Channel, but they came to nothing. The first serious attempt at digging a rail tunnel was promoted by a Victorian entrepreneur, Sir Edward Watkin, in 1881. Exploratory tunnels were dug at Dover in England and Sangatte in France, but work was abandoned in 1882, largely due to political pressure from the still-unconvinced British.

By the early 20th century, more advanced tunneling machines and the development of electric traction had made the idea of a rail tunnel a more realistic proposition. However, by this time a fit of xenophobic hysteria had seized most of the British military establishment, which was adamant that foreign powers would use the tunnel as a corridor for invasion. In fact, Marshal Foch, the Allies' Supreme Commander in the last year of World War I, declared that a tunnel could have shortened the conflict by two years. Nevertheless, these delays in building a tunnel simply meant that the competition kept on speeding up. By 1852, the

EARLY DESIGNS
This 1876 engraving shows plans for a single-track rail tunnel under the English Channel.

journey time between London and Paris had been reduced to 12 hours by sea and rail, and 60 years later it took a mere 7 hours. In the 1930s, the glamorous *Flèche d'Or* (Golden Arrow) train-ferry service reduced the journey time to just over six and half hours and by that time there was also a regular air service between Croydon, just outside London, and Le Bourget, northeast of Paris.

Although military hysteria dissipated after the end of World War II in 1945, progress on a tunnel remained slow, particularly at the British end. In 1963, the British government finally endorsed the building of a tunnel. By this time, the commercial outlook was promising as the number of travelers from London to Paris was increasing steadily (1 million per year in 1960 and 2.5 million by 1978). However, the British were still rather lukewarm about the project and had not yet agreed to a new rail line to link the tunnel to London and beyond. British Rail suggested a variety of routes but each one was more expensive than its predecessor. In the end, the cost of the new rail link provided an excellent excuse for the government to cancel the project in early 1975, just as tunneling was about to begin. This about-turn, which was in fact necessitated by Britain's perilous financial situation, naturally infuriated the French and reinforced their suspicions that Britain was not really serious about a Channel tunnel.

However, the idea for a tunnel refused to go away, and within four years Sir Peter Parker, the chairman of British Rail, together with his counterparts from the SNCF in France, had resurrected the project by proposing a single-track tunnel—inevitably called the "mousehole"—running "flights" of trains back and forth across the Channel. This somewhat ramshackle idea did not progress, but the principle of building a Channel tunnel received a boost from an unlikely source—eurosceptic British Prime Minister Margaret Thatcher. The "Iron Lady" decided that a tunnel could proceed under the Channel, as long as it was privately, not publicly, funded. Along with French President François Mitterrand, Thatcher set up a working group and then invited bidders to submit their proposals. Thatcher disliked railroads, so her natural preference was for a road-based, or at least a "drive-through" concept—a bias she shared with Mitterrand—but in the end an idea involving dual rail tunnels won out. By the end of 1985, the Channel Tunnel Group/TransManche Link, a consortium of five French and five British contractors plus five banks, had been awarded the contract to build the tunnel.

The whole operation was still fiendishly complicated. It took two more years for the Channel Tunnel Act, which officially rubber-stamped the program, to pass through the British Parliament (although the French procedure took mere days) and the world's investment community was by no means enthusiastic about the tunnel's financial prospects. Financial negotiations were hampered by the geographical spread of the various banks involved and the sheer number of contractors involved in the program made the construction process equally complex. However, the balance of power between the interested parties was transformed in early 1987 with the appointment of Alastair Morton as full-time British joint chairman of Eurotunnel, the company that actually held the contract to build the tunnel. During Morton's nine years at Eurotunnel he saved what was a fundamentally uneconomic venture from disaster by dealing with the British government, the contractors, and the 200 banks eventually involved. He also grappled with British Rail and SNCF, who would be the main users (to add further complexity, British Rail was being prepared for privatization at the time). After protracted negotiations about the technology the tunnel would use and the type of links the tunnel would provide, a complicated international consortium was formed combining railroad companies from France, Britain, and Belgium.

DIGGING UNDER THE SEA
Constructing the three tunnels required the use of 11 enormous tunnel boring machines (TBMs), each more than 492ft (150m) long.

Tunneling commenced at long last in December 1987, when the British started digging a service tunnel that would be used for maintenance and emergencies. Starting with the service tunnel was a good way of testing the tunneling conditions, foreshadowing any geological or practical issues before work on the main rail tunnels began. The French commenced their end of the service tunnel a couple of months after the British and then, in June 1988, two rail tunnels were started simultaneously from both sides. Thus the Channel Tunnel was, and remains, in fact three tunnels—two rail tunnels separated by a service tunnel. It was the biggest engineering project ever undertaken by France or Britain, requiring 15,000 workers at its peak. However, progress was slower than predicted, causing costs to spiral. In fact, there were no fewer than three major financial crises while the tunnel was under construction and the cost of the project exceeded its budget by more than 80 percent. Nevertheless, on December 4, 1990, the two tunneling teams finally shook hands mid-Channel. Amazingly, the tunnels were only 13in (330mm) out of alignment, despite each traveling 9 miles (15km) from opposite ends.

Completing the tunnels was just the beginning of the process, however. As one observer put it: "turning the tunnels into a piece of complex, safe, and sophisticated transportation infrastructure was to

prove an altogether different challenge," involving as it did power supply, lighting, ventilation, communication, maintenance provision, and fire detection and suppression equipment. The trains themselves were also incredibly complex, as they had to cope with different power supplies and signaling systems not just in Britain and France, but also in Belgium, where some trains would terminate. Safety fears also meant that every piece of equipment had to receive a separate operating certificate for each country, at an estimated cost of over $645 million, with a further $322 million lost in the resulting delays. However, the stringent safety requirements have proved their worth: there have been only three serious fires in the Channel Tunnel—all started on shuttles carrying trucks—that closed the Tunnel for short periods, but caused no loss of life.

In the end the tunnel was delayed by a year, far less than many much smaller projects. The contractors handed over the project to Eurotunnel in December 1993, and the "Chunnel"—as it was nicknamed—was officially opened by Queen Elizabeth II and President Mitterand in May 1994. Freight trains began traveling through the Tunnel a month later and rail passengers were welcomed in November, on the service now known as Eurostar. On December 22, 1994, the Channel Tunnel achieved a world first with the inaugural car shuttle service, known as Eurotunnel Le Shuttle, from Folkestone to Calais.

However, in one major respect the project was incomplete when the Channel Tunnel opened in 1994. Providing a fast line to the tunnel was relatively easy for the French, since all it took was building a branch line to Calais from their new fast line from Paris to Lille and Brussels. By contrast, neither British Rail nor the British government could overcome the difficulties inherent in building an entirely new line from London to Folkestone. British Rail had never designed a new railroad line, and all the routes it suggested threatened the rich heartland of Kent, as well as numerous marginal constituencies in South East London. As a result, the Eurostar trains on the British side of the tunnel had to travel along tortuous, winding tracks that had been laid a

NUMBER OF TONS OF SPOIL REMOVED DURING TUNNELING

11,023,000
(10,000,000 TONNES)

EUROSTAR
When the Channel Tunnel opened, trains on the French side could travel up to 186mph (300kph). On the British side, the top speed was only 100mph (160kph).

century earlier. The rails were strengthened and the route was provided with new signals and upgraded power supplies, but still, the trains were noticeably much slower in England than in France.

It took until 1996 before work began on a high-speed line for the British part of the route—a line now called High Speed 1 (HS1). Even when work finally began, it was far from straightforward. The project was beset with financial and construction problems. Excavation of the route revealed many ancient relics, which thrilled archaeologists, but caused further delays. Eventually, the first section of the high-speed route through Kent opened in 2003, reducing the journey time between London and Paris by 20 minutes.

The second section of the route, from the North Downs in Kent to a new terminal at St. Pancras station in London, was an even bigger engineering problem. It involved tunneling under the Thames River, navigating marshes in east London, building an enormous new station at Stratford, and rebuilding St. Pancras station. The biggest headache, however, was constructing an 11-mile (18-km) tunnel to St. Pancras that would avoid the existing sewers, main railway lines, and underground system. Finally, on November 13, 2007, the link was finished, and Eurostar transferred its passenger services from Waterloo to St. Pancras. The switch of terminals marked an historic moment, linking Britain's first high-speed line to its equivalent in Europe and finally completing the Channel Tunnel project more than 200 years after it had first been proposed. It is the world's longest undersea rail tunnel—23½ miles (37.9km) of the 31⅜-mile (50.4-km) tunnel runs under the Channel—and has been described by the American Society of Civil Engineers as "One of the Seven Wonders of the Modern World."

Building Tunnels

Tunneling is the most costly and labor-intensive of all engineering enterprises and, in the early days of the railroads, it was also the most dangerous. Working long shifts lit by candlelight in cramped conditions, workers risked serious injury or loss of life as they burrowed underground using only basic tools. From the mid-19th century their picks, hand-drills, and explosives were gradually replaced by tunnel-boring machines (TBMs), the first of which was used after 1862 to dig the Fréjus Rail Tunnel beneath the Alps (see p.106). Since then, strict safety regulations have reduced risks considerably, and the introduction of computer control has increased the machines' efficiency.

The tunnel's girth is determined by the cutting wheel's diameter— 26–29ft (8–9m) for the Channel Tunnel

Immersed tunnels

Suitable for routes that cross shallow bodies of water, immersed tunnels on the seabed are a cost-effective alternative to boring beneath it. The first tunnel to be built in this way was the Bay Area Rapid Transit (BART) tunnel in San Francisco in the late 1960s. Sections of the tunnel are floated to the tunnel site, sunk into a precut trench on the seabed, and secured with layers of gravel, concrete, and backfill.

BART TUNNEL SECTION UNDER CONSTRUCTION

TUNNEL-BORING MACHINE (TBM)
CHANNEL TUNNEL, c.1990s

Also known as "moles," TBMs are equipped with different cutting devices according to the geology of the tunneling area, from soft clay to mixed earth and shale or bare rock. The Channel Tunnel TBMs that cut through chalk marl under the English Channel to link Britain and France (see pp.348–53) were also designed to deal with high water pressure.

The cutting wheel rotates as hydraulic rams drive the machine forward

The 11 Channel Tunnel TBMs—each weighing 1,210 tons (1,100 tonnes)—were assembled on site

How it works

TBMs consist of several connected systems that are operated in different phases of construction to bore and line the tunnel, lay the rails, and convey waste material away from the site. The cutting wheel can be up to 63ft (19.3m) in diameter, while the TBM apparatus can be as much as 490ft (150m) in length.

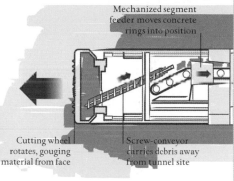

Mechanized segment feeder moves concrete rings into position

Cutting wheel rotates, gouging material from face

Screw-conveyor carries debris away from tunnel site

STEP 1: TUNNELING PHASE
During the tunneling phase, the rotating cutting wheel is pressed into the tunnel face at a predetermined rate. Debris is transported away from the cutting face by a screw conveyor (for shale or rock, pictured here) or a series of pressurized pipes (for earth or clay).

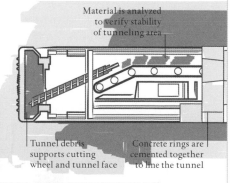

Material is analyzed to verify stability of tunneling area

Tunnel debris supports cutting wheel and tunnel face

Concrete rings are cemented together to line the tunnel

STEP 2: RING-BUILDING PHASE
The cutting apparatus stops to allow the ring-building phase to commence —the installation of concrete rings, which provide a watertight and strengthening lining for the tunnel. The rings are cast above ground and moved to the construction site via rail.

Switzerland: the Best of the Best

SWITZERLAND JOINED THE RAILROAD AGE relatively late, but, as in other countries such as Germany, the railroads quickly became a unifying force and a vital part of the nation's prosperity. So much so that today Switzerland can lay claim to having one of the most efficient and well-used railroad networks in the world.

Surrounded by mountains and beset by cold temperatures and copious snowfall in the winter, Switzerland did not seem like fertile territory for the iron road. Moreover, it did not even become a unified nation until 1848, following a brief civil war between the Protestant and Catholic cantons (states). The new Switzerland adopted a federal state system, similar to that of the US, and the new government decided that the country needed a proper, planned railroad system. So, in 1850, two British engineers, Robert Stephenson and Henry Swinburne, were asked to oversee the development of a rail network. The pair suggested a basic, east-to-west line along the valleys between Geneva and Zürich, plus another line from Basel to Lucerne. Rather oddly though, the capital, Bern, was to be on a branch line. The cantons built these early lines, often with foreign investment and contractors. Stephenson and Swinburne's influence also meant that the Swiss adopted the standard gauge (4ft 8½in/1,435mm) and instituted British-style left-hand running on double-track lines, both of which are still used today.

The iron road's progress in Switzerland remained slow. Unlike most European countries, where the demands of traditional heavy industries such as mining or agriculture fueled railroad expansion, it was tourism that eventually became the catalyst for further development in Switzerland. From the mid-19th century, increasing numbers of visitors flocked to resorts in the scenic Alps and lakes. However, when pioneering travel agent Thomas Cook took his first tour party to Switzerland in 1863, there was still barely 400 miles (650km) of railroad in the whole country. It was clear that the country's existing rail network could not cope with the demands of the tourist industry.

So, by the last third of the 19th century, the Swiss federal government was eager to see its railroads develop more rapidly. Furthermore, the country's location in the center of Europe made an efficient railway network in Switzerland highly desirable for its neighbors, too. They were eager to traverse Swiss territory to create both passenger and freight links across Europe, and were happy to invest in Swiss railroads to make that possible. A treaty between Switzerland, Germany, and Italy in 1869 agreed to a strategic link under the Gotthard Pass in the Alps. Completed in 1882, the Gotthard Tunnel (see pp.106–107) linked not just the cantons of Uri and Ticino, but also provided a valuable north–south route between Germany and Italy; the Simplon Tunnel (see pp.106–107), opened in 1906, also linked Switzerland to Italy.

Meanwhile, the railroads continued to spread across the country, including Europe's first rack railroad (see pp.108–109), which opened in 1871. Built to take tourists up the Rigi Mountain above Lake Lucerne, it used a rack-and-pinion system developed by Swiss engineer Niklaus Riggenbach. Three years later, the first 3ft 3⅜in- (1m-) gauge line opened from Lausanne to Bercher via Echallens, although this was actually built to serve an agricultural area rather than the tourist industry. By adopting a narrower gauge in less accessible or less populous areas, lines could be built much more economically, allowing railroads (and also light rail) to spread into the steepest and most remote corners of Switzerland.

SPANISH BUN RAILROAD

The first Swiss railroad linked Zurich and Baden in 1847. It was nicknamed *Spanischbrötlibahn* ("Spanish Bun Railroad") because affluent Zurich bankers used it to send their servants to fetch Baden's specialty buns.

GOTTHARD LINE
The trans-Alpine Gotthard Railway proudly became electrified in 1922.

After the initial railroad boom of the late 19th century, some speculative railroads went bust and many Swiss people became disenchanted with their railroads being largely owned and operated for the benefit of foreign shareholders. Consequently, in 1898 a referendum on nationalizing the railroads led to the formation of the Swiss Federal Railways in 1902 (officially known as SBB-CFF-FFS). Over the next few years, the SBB-CFF-FFS acquired approximately 50 percent of the national trackage and embarked on a program to electrify the whole network. Given the country's lack of coal, the Swiss were early adopters of electric traction: the first electrically powered railroad in Switzerland was a 3-ft 3⅜-in (1-m) gauge line between Montreux and Chillon in 1888. The Simplon Tunnel was also electrified from the start, and electric traction was used, too, for the *Jungfraubahn*, completed in 1912.

The two world wars accelerated the electrification process in Switzerland: during World War I coal shortages caused widespread travel disruption, while World War II affected Switzerland even more, despite its neutrality. The tourist trade on which many railroads depended dried up, and some lines closed in the aftermath of the wars. However, the postwar government soon realized the value of Switzerland's transportation infrastructure and began to invest in the railroads, so much so that by 1960 the entire railroad network in Switzerland was electric, making it the first country in the world to achieve this milestone.

During the latter half of the 20th century, car and semitruck traffic in Switzerland grew inexorably, as it did elsewhere. With the building of modern highways through and under the Alps, much freight that was once rail-borne transferred to the road. By the 1980s, however, the Swiss public was increasingly concerned about the environmental damage caused by heavy semitrucks. This eventually led to two major national referendums on the railroads. The first, in 1987, led to the Bahn 2000 project, which aimed to bring the Swiss rail network into the 21st century. This initiative resulted in the improvement of many lines and the construction of sections of high-speed track that are an integrated part of the existing infrastructure rather than, as in many other countries, entirely separate. A second referendum in 1992 approved, by a

IN THE SHADOW OF THE ALPS
The Matterhorn-Gotthard-Bahn (MGB)
is a narrow-gauge Alpine railroad. This
branch line at Andermatt links to the
northern end of the Gotthard rail tunnel.

massive majority, a project called "AlpTransit" that regulated the number and capacity of semitrucks allowed across the Alps. It prioritized improvements to the rail infrastructure over road-building, intending freight to be transferred onto the trans-Alpine railroads. To facilitate this plan two new railroad tunnels were proposed—the 21-mile (34-km) Lötschberg Base Tunnel, which opened in 2007, and the 38-mile (61-km) Gotthard Base Tunnel, which is due to be completed 2016.

The 1980s also saw the introduction of an impressive integrated national transportation schedule in Switzerland, coordinating rail, bus, and trolley services. It means that almost every station in Switzerland has at least an hourly rail service that links to a national timetable plan, giving regular connections with all other routes and modes of transportation. Oddly, although the transportation services are coordinated by the government, its national rail company, SBB-CFF-FFS, owns just under 2,000 miles (3,200km) of track, with the remaining 1,200 miles (750km) being operated by around 80 different private or seminationalized companies. Despite this, the Swiss railroad system is far more streamlined and has far less duplication than many other networks around the world, where competition between different companies resulted in unnecessary or inefficient lines being built.

Historically, the Swiss electorate, and generally their politicians, have been far more conscious of the implications of their public transportation than most other nations. For example, the standard gauge rail network is the core of a wider transportation system in Switzerland into which local trains (often 3-ft 3⅜-in gauge/1-m), trolleys, and buses link at all stations. Despite this integrated network, the Swiss still have one of the highest rates of automobile ownership per capita (about one for every two people), but they use them far less. In fact, Swiss people make more journeys via railroad than any other nation, apart from Japan. On average, the Swiss travel 1,060 miles (1,706km) per person annually by rail, while the Japanese travel 1,200 miles (1,900km). (The Swiss actually make more individual journeys than the Japanese, but these journeys are shorter.) Given that Japan is a far bigger

THE HEIGHT ABOVE SEA LEVEL OF JUNGFRAUJOCH STATION IN THE ALPS

11,332FT
(3,454 M)

THE HIGHEST IN EUROPE

THE SWISS RAIL NETWORK

country with a very dense population, Switzerland's record is all the more remarkable. The key to that success is that the rail network is part of a genuinely integrated public transportation system designed to make nonmotorized travel easier and more efficient for its citizens. Reliability and punctuality have proved crucial, too, supported by consistent government investment in the railroads. Not only is the rail network extensive and well managed, but the cost of rail travel is also highly competitive. In addition to the usual range of discounted season and zonal tickets, it is possible to purchase a national, all-modes (rail, bus, trolley, boat) travel card that costs around $5,630 (£3,500) per year—a bargain by the standards of many other countries.

Today, a growing Swiss population is placing the rail network under increasing strain and the railroads require considerable investment to keep pace with demand. The Swiss, however, seem up to the task: in Zürich, for example, an expensive series of tunnels has been planned to allow trains to pass through the city more quickly and to link expanding suburbs. Railroads, it seems, have become ingrained in the culture and success of Switzerland, and the importance of the iron road shows no signs of waning.

SIGNATURE STRUCTURE
The Glacier Express crosses the imposing
Landwasser Viaduct, in Graubunden,
Switzerland, along the Albula railroad.
The limestone bridge rises 213ft (65m)
above the Landwasser River.

Going Faster: Bullet Trains and High-speed Lines

B Y THE 1960s, THE RAILROADS were competing against automobiles, semitrucks, and airplanes. Politicians and civil servants thought the railroad age was coming to an end—an outmoded form of travel that belonged to the 19th rather than the late 20th century—and governments chose to invest in highways and roads that provided the convenience of door-to-door travel instead. Car ownership soared, semitrucks started hauling freight, and commercial flights took off, heralding the dawn of the jet age. To counter these trends, railroad services needed modernization. It was the Japanese who led the way with the groundbreaking *Shinkansen* ("new mainline"), known in the West as the bullet train. It started a trend for high-speed railroads that would spread—albeit rather slowly—around the world.

Japan's geography and pattern of settlement contributed to the genesis of the bullet train, as it had to the development of Japan's railroads in the first place. The country consists of four main islands, but less than a fifth of the land mass is habitable, so most of the population of 126 million is confined to a relatively small part of the country. It was this density of population in lowland regions that created the right conditions for high-speed rail development. In fact, while the high-speed aspect of the *Shinkansen* is emphasized, the new lines evolved mainly because the existing network was running out of capacity.

Japan came late to the railroad age. The island state had shunned the rest of the world until 1868, when the new Meiji administration sought to modernize the country, opening its doors to the railroads. The first line, which covered the 18 miles (29km) between Tokyo and Yokohoma, opened in 1872, and the Japanese adopted trains with enthusiasm. The engineers of the line used the narrow gauge of 3ft 6in (1,067mm), the same as in New Zealand as the islands had similar topographies. Later plans to convert to standard gauge to increase traffic on the railroads were opposed by military powers; instead, lines were extended.

The railroads boomed in the final few years of the 19th century, and by 1907 there were nearly 4,500 miles (7,250km) of railroad in operation, all owned by the state-run Japanese Imperial Railways. Development continued steadily between the first and second world wars. By 1945, the

INAUGURATION OF THE TOKAIDO *SHINKANSEN*
The Tokaido *Shinkansen* was launched in Tokyo just in
time for the 1964 Olympics. The bullet train reduced the
journey time between Tokyo and Osaka to 4 hours and
became the prototype for high-speed trains worldwide.

total mileage had reached 16,000 miles (26,000km), nearly a quarter of
which was owned privately, with the rest under the control of the
renamed Japanese National Railways.

By the 1930s, Japan's first trunk route, the Tokaido line, which linked
Tokyo with a series of major cities that included Nagoya, Kyoto, Osaka,
and Kobe, was already becoming heavily congested. An entirely new
route between Tokyo and Osaka was proposed that would cover the 400-
mile (640-km) journey in four hours—an unheard-of speed for a
railroad at the time. Work started in 1941, but was more or less abandoned
after the Japanese attack on Pearl Harbor triggered the Pacific War.

The Japanese economy took some time to recover from defeat in the
war, but by the mid-1950s the Tokaido line was again running at capacity
and the plan for the Tokaido *Shinkansen* was revived. Shinji Sogō, the
president of the state-run railroads, lobbied hard to persuade the
government that the railroad line would be viable, as at the time it
seemed that cars and planes would make railroads redundant. It was the
need for extra train routes that stimulated the *Shinkansen*'s development,
but to compete with cars it was built on highway principles, with few
stops and fast journey times. The high speed was a by-product of the
need for capacity rather than an end in itself—a principle that applied to
most high-speed systems around the world. Of course, the high-speed

PASSING MOUNT FUJI
In what has become an iconic image of the meeting of the modern and the ancient world in Japan, the *Shinkansen* bullet train passes in front of snow-capped Mount Fuji.

aspect helped to attract passengers away from their cars, and also made it possible for rail to compete with aviation over distances of up to 500 miles (800km), thanks to its city-center-to-city-center routing.

It was decided that Japan's new line would carry only fast electric passenger trains and would use the standard 4ft 8½in (1,435mm) gauge to provide greater capacity and enable use of technology from other railroads. Despite difficult terrain and cost overruns—the eventual cost of 380 billion yen (around $1.1 billion at the time) was twice the original budget—the line was completed five years after work started in 1959. By contemporary high-speed standards, the line was slow, with an average speed of 130mph (209kph), but thanks to its dedicated track and limited stops the journey time was radically reduced. While the conventional express took 6 hours and 40 minutes between the two main cities, Tokyo and Osaka, the *Shinkansen* made the trip in 4 hours when it started operating just before the 1964 Tokyo Olympics. The *Shinkansen* changed business patterns too, by making day trips between Tokyo and Osaka possible. Shinji Sogō's faith in the railroad was fully justified. An immediate success, the service carried 100 million passengers in less than three years; by 1976, it had carried one billion. Today, the Tokaido *Shinkansen* carries 143 million passengers annually.

It was not an easy ride, however. The service suffered teething pains, including discomfort to passengers' ears when trains crossed in the tunnels that accounted for 45 miles (72km) of the route, and a more embarrassing problem: the air currents generated in the tunnels blew water up from the toilet bowls, much to the users' discomfiture. Eventually, it was decided to pressurize the trains to solve these problems, which proved expensive but was successful. Interestingly, despite the *Shinkansen*'s popularity, Japanese National Railways faced opposition to building the network it had envisaged. Noise and cost were both concerns, but eventually, in the 1970s, several new lines as well as extensions to the Tokaido line were built. Today, there is a network of 2,200 miles (3,540km). The fastest trains on the Sanyo line travel at 186mph (300kph), a speed that has become the norm across the world for high-speed lines.

NUMBER OF PASSENGERS CARRIED BY BULLET TRAINS EACH YEAR

325 MILLION

JAPANESE HIGH-SPEED RAIL NETWORK

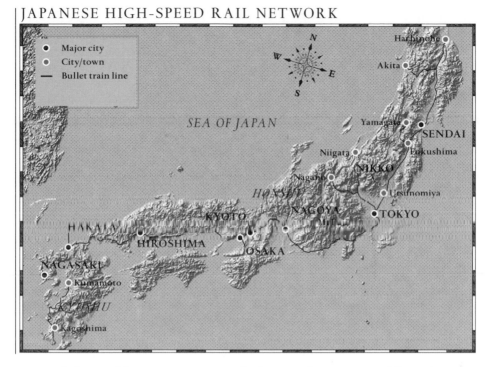

Around the world, it was some years before another country followed Japan's lead. Although railroad managers were eager to increase speeds from the 60–70mph (90–110kph) that was standard for express trains by the 1960s, governments remained doubtful as to whether the "old technology" of train travel was worth investing in. In Germany and France, train speed trials took place, demonstrating that trains could easily reach speeds of 124mph (200kph) for long periods, and attempts were made to speed up services on existing lines. In France, the Paris–Toulouse route was upgraded in 1966 to support 124mph (200kph) running through improvements to tracks and signaling. In Britain, a new diesel service branded InterCity 125 was introduced on routes from London after 1976. However, all used existing lines, which limited their speed as other, slower trains shared the tracks.

To avoid line-sharing, France decided in the 1970s to create a dedicated bullet-style service called the *Train à Grande Vitesse* (TGV); like the *Shinkansen*, the TGV has become a world-renowned brand. At the time, capacity on the key Paris–Lyon route was reaching its limits and it was decided to build a new line, separate from the existing railroad except at the city entrances, where tracks were shared with conventional services. The

TGV uses the same gauge as regular trains so it can run on high-speed and standard tracks for flexibility. The French had created high-speed trains that had beaten the world record several times, reaching an impressive 262mph (422kph) on an experimental track in 1969. Clearly, such speeds would only be possible on a dedicated track, so in 1976, construction of the first *Ligne à Grand Vitesse* started. The Paris–Lyon line was completed in 1981, when the first TGVs started running at a top speed of 168 mph (270kph), later increased to 186mph (300kph). The service was an instant success, challenging air travel between the two cities. France embarked on a network of high-speed lines emanating from Paris. The *Est*, which opened in 2007, operates at 200mph (320kph).

France's high-speed network covered 1,185 miles (1,907km) in 2013, but has now been surpassed in distance by Spain. In 2005, Spain announced a plan to ensure that 90 percent of the population would live within 30 miles (50km) of a station served by the Spanish *Alta Velocidad Española* (AVE) network. Spain's first high-speed line, between Madrid and Seville, was completed in time for the Seville Expo '92. It used standard gauge as opposed to the 5-ft 6in (1,680mm) Iberian gauge used by other Spanish services. This led to a technological development that allows the latest high-speed trains to change gauge without stopping so they can continue journeys off the high-speed network. By 2013, Spain operated half a dozen routes out of Madrid and, with Barcelona also a hub, a total of 1,285 miles (2,000km) of high-speed line had been opened, with a further 684 miles (1,100km) under construction.

After France, Germany was the second country in Europe to develop high-speed lines. It opted for a different model by creating high-speed sections rather than whole new routes, so the trains switch frequently between conventional and high-speed lines. Germany launched its Intercity-Express (ICE) in 1991, operating at a top speed of 174mph (280kph) on the Hannover–Würzburg high-speed railroad.

Since 2000, a number of new high-speed rail services have started operating in East Asia. In South Korea, the Korea Train Express was launched in 2004 along the Seoul–Busan corridor, linking the country's two biggest towns. It uses trains built by Alstom, the same company that builds the TGV in France. Taiwan has a high-speed railroad running more than 200 miles (322km) along the west coast of Taiwan, from the national capital Taipei to the southern city of Kaohsiung, using technology based primarily on Japan's *Shinkansen*. The biggest network by far, however, is in China (see pp.372–79).

TGV AND THALYS HIGH-SPEED TRAINS
Connecting France, Germany, and Benelux, TGV
and Thalys trains depart from stations in Paris
before joining dedicated high-speed lines.

Safety has been a major factor in the success of high-speed trains: although there have been three major accidents involving high-speed trains, none occurred at full speed on dedicated lines. In Germany in 1998, a wheel broke at 124mph (200kph) and came off the rails at a bridge resulting in the derailment and destruction of the full set of 16 cars and the death of 101 people. In China in 2011, due to a signaling error, a train traveling at 62mph (100kph) hit a stationary train on a viaduct, killing 40 people. At Santiago de Compostela in Spain in 2013, a train came off the rails at 120mph (195kph) on a curve with a speed limit of 50mph (80kph), and smashed into a concrete wall, killing 79 people. Despite these accidents, the overall safety record of high-speed rail is better than for any other form of transportation.

By 2013, there were nearly 10,000 miles (16,000km) of high-speed line around the world, with plans for lines in countries including Ukraine, Turkey, and Belarus. In the US, an 520-mile (837-km) line from San Francisco to Los Angeles has been approved by the California state government, but budget constraints and protests have delayed construction. In Britain, the proposal for HS2, a 330-mile (530-km) high-speed line network joining London with several major cities at a cost of $82 billion has also met with fierce opposition. While some new plans might be tinged with controversy, the advantages of high-speed rail are uncontested, and its place in the future of transportation, is assured.

China, the New Pioneer

CHINA WAS PROBABLY THE ONLY COUNTRY in the world to have joined the railroad age and then, briefly, to have left it again. Despite this false start, China now boasts by far the most extensive system of high-speed dedicated lines anywhere on the globe, a network that China is bent on expanding to become the backbone of the nation's transportation infrastructure. China is also home to the world's highest railroad: the Qinghai–Tibet Railroad.

The first railroad line in China was built by Jardine, Matheson & Co., a European-owned trading company looking to improve access between Shanghai and the nearby port of Woosung. The line was just 10 miles (16km) long, but its construction in 1876 was mired in controversy: the deeply conservative Chinese officialdom was reluctant to allow the laying of railroads, fearing it would ruin the livelihood of the vast numbers of those who carried goods for a living. One Chinese official at the time, Yü Lien-yuan, worried that the resulting unemployment would foment unrest:

> several tens of millions, who earn their living by holding the whip or grasping the tiller, will lose their jobs. If they don't end up starving in the ditches, they will surely gather [as outlaws] in the forests.

Another official was concerned that coal would run out, arguing that "when one uses coal with such profligacy, coalfields would soon disappear." In addition to these doom-laden visions, there was much antagonism toward foreigners and foreign-owned concerns at this time. Imperialist shows of might, such as the Opium Wars, lived in very recent memory, and foreign powers had been taking advantage of China's weakness: several European governments, as well as the Japanese, had set up missions along the coast that were effectively a way of obtaining access to China's riches without paying taxes on them. As such, Chinese officials took a dim view of the Shanghai–Woosung Railroad, and its European-sponsored construction never received official sanction. Just one year after the railroad opened, Shen Pao-Chen, the governor of the region through which the railroad ran, ordered the line to be ripped up and had the equipment shipped to Taiwan, where it was abandoned to the elements.

THE WORLD'S HIGHEST RAILROAD
A train passes through the Kunlun
Mountains on the Qinghai–Tibet Railroad,
the world's highest rail line, which reaches
16,640ft (5,072m) above sea level.

It was not until 1881 that a permanent railroad line would open in China. Originally intended to be mule-hauled, the line was a 6-mile (10-km) standard-gauge railroad running from a coal mine to a canal at Hsukochuang, about 100 miles (160km) east of the capital, Beijing. A British engineer, C. W. Kinder, was responsible for the construction, as well as for commissioning China's first locomotive, the *Rocket of China*. These events, however, did not herald a railroad revolution. The government remained reluctant to endorse this groundbreaking method of transportation, despite its success across the world, and very few lines were built in the 1880s. It took the disastrous defeat in the Sino-French War of 1884–85 to make the Chinese realize that modernization was essential and that railroads could be a catalyst for development. Kinder's line was extended by 20 miles (32km) in the direction of Beijing. However, in what can be seen as a measure of the role of superstition in Chinese government structures, a mysterious fire in the Imperial Palace was seen as a sign of celestial displeasure and the line was never completed.

By 1894, when the Sino-Japanese war broke out, little progress had been made and China had a mere 320 miles (500km) of railroad, compared with 175,000 miles (280,000km) in the US. However, defeat in the war finally stimulated a railroad boom in China. While Beijing became the center of the network, many other lines were built to serve mines in relatively remote areas. By the time of the Xinhai Revolution of 1911, which created the Republic of China, there were 6,000 miles (9,500km) of track, a significant increase, but still a relatively small statistic for the most populous country in the world. It was, at the time, half the size of the railroad network in India—a similarly impoverished, but smaller, nation.

Growth of the railroads slowed during the period of the Republic as a result of a series of civil wars and the occupation of China by Japan in the late 1930s. Many lines were destroyed in these various conflicts, and by the end of World War II this vast nation still had only 14,000 miles (22,500km) of workable railroad. After gaining control of the country for the communists in 1949, Mao Zedong invested heavily in the railroad network. Lines were repaired and new ones built, even in difficult mountainous territory. This progress continued after Mao's death in 1976, and by the end of the 20th century China finally had a network covering most of the country. One major gap, however, remained—a line connecting Tibet with the rest of China.

Tibet is remote and separated from the rest of the country by the Kunlun Mountains in the north and the Nyenchen Tanglha range in the east. The main Tibetan plateau is a huge, high landmass stretching 1,500 miles (2,400km) from east to west, and 500 miles (800km) from north to south. It is home to the largest subarctic permafrost region in the world—which, to put it mildly, is not ideal railroad territory. All land routes to the vast plateau cross mountain passes that climb higher than any peak in the US. As author Abrahm Lustgarten describes in *China's Great Train*, the roads

> twist and wind through steep gorges loaded like cannons with unstable rock and snow at their peaks and flushing with torrents of interminable water in their troughs

A China–Tibet rail link was seen as a way to cement China's control over this long-disputed area, known by the Chinese as the Tibetan Autonomous Region. Historically part of China, Tibet had declared its independence following the collapse of the Qing dynasty in 1912, but had been reclaimed by the communist-controlled government in 1951 and occupied by soldiers from the People's Liberation Army. Ever since the communist revolution, the government had harbored

APPROACHING LHASA
A train charges along the 3,047-ft (929-m)-long Lhasa River Railway Bridge, 1¼ miles (2km) from the terminus of the Qinghai–Tibet Railway.

WORKER'S BADGE
This badge displays the insignia of the China Railway Corporation, which employs some 2 million people.

ambitions to build a railroad line to Lhasa, the capital of Tibet, to help establish control over the territory. However, technical difficulties and lack of money had stood in the way of the project. International experts argued that the railroad simply could not be built, having observed the difficulties of laying railroad tracks on permafrost during the construction of the Baikal Amur Mainline in eastern Russia (see pp.330–37).

Tibet accounts for an eighth of China's landmass and yet in 2000 it was still the only region in the country without a rail link connecting it to the more developed east. The Chinese government instituted a "Go West" campaign and the proposed line to Tibet became an important part of that strategy. At the time, Tibet was an undeveloped agrarian region with little connection to the outside world, but there was a possibility that mineral resources could be exploited with the help of the proposed railroad.

A precursor to the Tibetan Railroad was completed in 1984: an 500-mile (800-km) railroad heading west from Xining (the capital of Qinghai province and the traditional gateway to Tibet) to Golmud, also in Qinghai province. The plan to continue the line through to Lhasa, however, would not be agreed to until over a decade later. In 1999, President Jiang Zemin launched a campaign to develop western China, which was lagging behind the booming east, and the Tibet railroad became a key part of that strategy. However, there remained some dispute over the best route for the railroad. Golmud, which had been founded in the 1960s as a labor camp for mainly Tibetan prisoners, was now a small city and offered the shortest route to Lhasa. This route involved crossing hundreds of miles of permafrost, however, and there were concerns that the technology to ensure that this could be done was not available. The most obvious alternative route was one from Yunnan province in southern China, but this would be twice the distance of the Golmud route, and so twice as expensive. Eventually, it was decided that it was possible to overcome the permafrost problem, and work started on the 710-mile (1143-km) line between Golmud and Lhasa in 2001. A railroad constructed in the 21st century benefitted from many previously unavailable techniques, but the difficulties of working at such high altitude in a remote region still surpassed that of most previous railroad projects, and the labor force required was enormous—over 100,000 workers migrated to Tibet to build the railroad at the start of the project. One major challenge was that some sections had to be built on ground that was not quite permafrost—the top layers of soil melted during the summer and so became unstable. To accommodate this, long sections of track were elevated on what were effectively bridges, held up by deep, pile-driven foundations. In addition, passive heat exchangers were installed to cool the track and the surrounding soil.

The human cost of the venture was high, with many workers succumbing to altitude sickness. According to Lustgarten, "Tibetans in nearby villages would see railway officials burying dead workers on the hillsides outside the [Fenghuoshan] tunnel", but the official explanation for the deaths was food poisoning. No casualty statistics have been released by the authorities, who deny that there were any deaths from altitude sickness.

The work began at both ends of the line, and track-laying was completed within four years—installation of signaling and other equipment took another year. Just five years after work started in July 2006, the line opened with much fanfare. The overall cost was around $4 billion—although this may be an underestimation given the difficulties of precisely working out the costs of the program. On its completion, the Qinghai Tibet Railroad beat numerous records in railroad construction. It is the highest railroad in the world: the Tanggula Pass, at 16,640ft (5,072m), surpasses its spectacular counterpart in the Andes, Peru (see pp.196–201), built almost a century before, by around 820ft (250m). Tanggula station is also the world's highest railroad station, and the ¾-mile (1.2-km)-long Fenghuoshan Tunnel is also the highest railroad tunnel in the world at 13,435ft (4,905m) above sea level.

The railroad has a capacity of up to eight passenger services per day in each direction. Since the air in Tibet is thin, the cars on Lhasa trains have special air conditioning systems to keep oxygen levels healthy, and each seat has its own emergency breathing apparatus. The windows are especially large to give travelers the best view, and are protected against the high levels of ultraviolet light on the Tibetan plateau. With all these dangers, passengers are required to obtain a Health Registration Card before traveling from Golmud to Lhasa, and each train has a doctor on board in case of emergencies.

China's belated railroad expansion has continued with the construction of a huge network of high-speed lines. Up until 1993, China's trains were still very slow, averaging just 30mph (48kph), prompting a number of "speed up" campaigns to counter competition from roads and aviation. The result was a series of services that could run at up to 100mph (161kph) by the end of the decade. The government, however, had even greater ambitions: the construction of a whole new set of lines in order to radically improve rail services and the nation's infrastructure. A program was thus devised to build the world's biggest network of high-speed lines, defined as more than 124mph (200kph). Some existing lines were upgraded, but for the most part entirely new lines were constructed—each one faster than the last. The "Mid-to-Long-Term Railway Network Plan" proposed the construction of a national, high-speed rail grid composed of four north-south corridors and four east-west corridors, which, together with upgraded existing lines, would total 7,500 miles (12,000km).

The first of these dedicated lines, the Qinhuangdao-Shenyang High-Speed Railway along the Liaoxi Corridor in the northeast, opened in 2003 with a line speed of 124mph (200kph), which was upgraded to 155mph (250kph) by 2007. Others soon followed, some opening in time for the 2008 Olympics. One of these was the Beijing–Tianjin Intercity Railway linking northern China's two largest cities and designed for trains running at a speed of 217mph (350kph).

In October 2010, China opened its fifteenth high-speed rail, the Shanghai–Hangzhou line, and the following year the key Beijing–Shanghai line, which had a design speed of 236mph (380kph), became operational. This gave China over 5,000 miles (8,000km) of dedicated high-speed track, more than double that of any other country.

These impressive advances received a setback in July 2011 with a disastrous crash at Wenzhou in which two high-speed trains derailed. The crash cast a

NUMBER OF BRIDGES ON THE GOLMUD-LHASA LINE

675

TOTALLING

100 MILES
(161 KM)

HIGH-SPEED ROLLING STOCK
A fleet of high-speed bullet trains awaits servicing
at a maintenance base in Wuhan, Hubei province,
one of China's busiest transportation hubs.

long shadow, and for a time construction slowed. The program threatened to be delayed, or even shelved, as passenger numbers dwindled in response to the accident, and the top speeds of trains were reduced. However, by 2012, the program had resumed and passenger numbers climbed again.

In 2013, China's 1,580 high-speed train services transported 1.3 million passengers daily, and these impressive statistics are only going to grow. The government plans to have a 16,000-mile (25,750-km), high-speed network by 2020—at a cost of $300 billion— giving China a high-speed rail system extending many times the distance of any other country. This achievement, along with the expanding number of subway lines in towns across the country— and the record-breaking Tibet Railroad—establishes China as indisputably the 21st century's principal railroad pioneer, so far.

ACROSS PLAINS AND MOUNTAINS
A train runs past a mountain range in the
Qinghai Province, northwest China, in 2006.
Although a late starter, China has become
the poster child for modern railroads.

The Railroad Renaissance

THE STORY OF THE RAILROAD would have been remarkable even if, with the advent of the automobile and plane, it had simply been consigned to history. In that, it would have resembled the trajectory of so many other innovations: birth, brief heyday, gradual decline, and then oblivion. However, despite a period of decline, the railroad has proved pessimistic predictions wrong by enjoying a 21st-century renaissance, and a future that looks assured.

This resurgence of the train has not simply been the result of far-reaching technological developments, although improvements have certainly been made. Railroad passenger cars are more comfortable than those of the 19th century, and freight cars are sturdier and better designed for rapid unloading with the spread of containerization. Signaling is much improved, too, and there have been other sophisticated adaptations to make the railroads faster and more efficient. Despite these advances, train travel in the 21st century would still be immediately recognizable to George Stephenson and other pioneers. Tracks are still usually set 4ft 8½in (1,435mm) apart, and passengers are still transported in cars that stop at stations and are controlled mostly by external signals.

Of course, the function of the train today is different. While the railroad was once a monopoly supplier of long-distance transportation, it is now more of a niche industry—but a very important one. Never again will the railroad serve every village and small town. The heyday of the iron road providing the only fast and cheap way of traveling between many places is over; those little village stations and rural halts are lost forever. Nor will railroads dominate the freight market as they once did. Gone, too, are the baggage rooms and carts once found at every station, wiped out by the semitruck and the van.

The railroads had to overcome a very difficult time: there was a point in the post-war period when railroads were regarded as irrelevant (see pp.340–47). The French railroad writer Clive Lamming even gave this phenomenon a name—"*ferropessimisme*," the notion that the decline and marginalization of the railroads was inevitable. But the railroads are still very much a part of modern life and pessimists have been proved wrong. Many countries are now bemoaning the fact that key lines were closed down and major stations turned into shopping malls or housing developments. Across the world, the railroads are flourishing and will continue to do so for the foreseeable future, because rail travel is still a very convenient form of travel for passengers and an extremely efficient way of transporting goods.

There are several key markets in which rail travel offers great advantages to passengers. Intercity rail journeys of 300–400 miles (480–645km)—or further with high-speed lines—may take longer than a flight, but passengers are able to relax or work on the train and, unlike far-flung airports, train terminals are located in the heart of the city. Local commutes are also more efficient by rail, whether train or subway, as these trips are usually faster and more reliable than driving, and unaffected by traffic. Trains also offer the best way to enjoy scenic routes, and in some cases, as with the Trans-Siberian (see pp.180–89), are one of the only viable means of transportation across remote areas.

Rail is also very well suited to moving large amounts of very heavy, nonurgent freight, such as gravel and stone, which otherwise damage roads. Trains have a competitive advantage too, when transporting loads long distances: they are cheaper than convoys of semitrucks, which require several drivers and may need to stop overnight. Developments in containerization have also made loading and unloading far easier.

The railroads have made a remarkable recovery from their post-war nadir, when it seemed their glory days were at an end and that the train would soon go the way of the schooner or the stagecoach. They survived, in most countries at any rate, because of their competitive edge in several markets, and because road transportation has its own limitations. In an age of uncertainty over oil supply, railroads remain a reliable and relatively cheap form of transportation.

The railroads, however, do not stand still. They are continually adapting and evolving—and expanding. In response to competition, services have been speeded up and facilities improved; redundant lines and services have been abandoned; and major developments can still be expected. One such example is the wider adoption of in-cab

A 1994 US STUDY SHOWED THAT SEMITRUCKS EMIT AROUND

8 TIMES

MORE AIR POLLUTION THAN TRAINS

signaling to replace external signals; this is safer and more efficient, allowing more trains on the track, but will require considerable investment. Moreover, and more significantly, brand-new and reopened lines are springing up in many places.

There are a huge number of significant projects in the works, many costing billions, or tens of billions, of dollars. Aside from China (see pp.372–81), Saudi Arabia is probably responsible for more major rail and subway developments than any other country. Its 900-mile (1500-km)-long North–South line will enable freight to travel from Al Jalamid (in the phosphate belt in the north) to Az Zabirah (in the bauxite belt in the center of the country), and then eastward to the processing and port facilities at Raz Az Zawr.

Also in the pipeline is the ambitious Landbridge Project, which is intended to run from the west to east coasts of Saudi Arabia, linking the Red Sea and the Persian Gulf, thus greatly reducing the time required to transport freight from the Gulf. It involves building 600 miles (960km) of new line to connect the capital city, Riyadh, to Jeddah on the west coast, while upgrading existing links between Riyadh and the city of Dammam on the east coast. In addition, 80 miles (130km) of track will run north–south along the coast to connect Dammam and Jubail. These links will create a useful new route for sending raw materials and manufactured goods between Europe and North America on one side, and East and South Asia on the other.

In addition to these bold initiatives, the Saudi government is building a Mecca-to-Medina line (the Haramain High Speed Rail) to carry pilgrims for the Haj, redolent of the Hejaz Railway (see pp.288–93). Other proposals include the Riyadh subway, to serve six routes, and the Saudi–Bahrain Railway Bridge.

All of these projects involve building lines across difficult desert terrain, often in remote areas, echoing many of the great construction projects covered in this book. It is clear that Saudi Arabia is embracing the railroads in a way that is astonishing for a country that is home to the world's largest oil reserves, with a wealth built on supplying gasoline for cars. There is no shortage of exciting forthcoming developments and projects worldwide. Russia is planning to improve the Trans-

THE PROJECTED COST OF
THE RIYADH SUBWAY IN
SAUDI ARABIA

$23 BILLION

VISIONS OF THE FUTURE
This digital projection of a terminus along the
Haramain High Speed Rail in Saudi Arabia shows
the platforms shielded by fabric roofing. The ambitious
designs for the main stations will enable massive capacity.

Siberian (see pp.180–89) and the Baikal Amur Mainline (see pp.330–37),
and has visions of building a line across to the United States. It would
certainly beat any other project for sheer ambition and, indeed, cost.

The renaissance of rail is arguably most remarkable in Africa—a
continent that has never properly exploited the advantages of rail
transportation. Now, thanks in many instances to investment from
China, several major lines have been brought back into use and others
are being constructed. A plan to build an 1,800-mile (2,900-km) line
across central Africa, linking the capital of landlocked Rwanda,
Kigali, with the Kenyan port of Mombasa, looks set to proceed with
Chinese money at a cost of $13.8 billion. The railroad will partly make
use of an existing colonial-era line in Kenya and Uganda, but the
section from the Ugandan capital, Kampala, through to Kigali will be
entirely new. Connections through to other parts of Kenya and
Uganda are also envisaged.

In West Africa, work is starting on a line in order to facilitate the
export of minerals. The line would link Niamey, the capital of
another landlocked nation, Niger, with the huge port of Abidjan in
Ivory Coast, via Ouagadougou, the capital of Burkina Faso. Nigeria
also has plans to renew and expand its railroads, and is seeking to
reinstate commuter services in its biggest city, Lagos. In South

Africa, a new 50-mile (80-km) mass rapid transit system, the Gautrain, was fully opened in 2012, linking Johannesburg, Pretoria, Ekurhuleni, and O. R. Tambo International Airport. There will never be a Cape to Cairo line (see pp.214–23), but Africa will become more rail-oriented than it has ever been.

The growth of the railroads is a worldwide phenomenon. In the fall of 2013, the railroad development website, railway-technology.com, listed nearly 450 major railroad projects across the world, including more than 100 in Asia and 33 in Australasia. As mentioned previously, there are plans for high-speed lines in numerous countries where previously rail investment had stagnated or declined. High-speed rail is a rising new market as it attracts travelers away from short-haul flights and onto the more environmentally sustainable railroads. It offers not only the prospect of reduced journey times between city centers, but also a far more pleasant travel experience.

Meanwhile, the other great boom in railroad development has been the subway, and this shows no sign of abating. In the summer of 2013, there were 188 subway systems across the world in no fewer than 54 countries, ranging from Teresina in Brazil to Yerevan in Armenia. Perhaps the most surprising adherents are to be found in car-obsessed

A FAILED ENTERPRISE
The underused, overpriced Sydney Monorail ran for 25 years, before being scrapped in 2013 to make way for a light rail system.

Dubai, the largest city in the United Arab Emirates, which opened a subway line in 2009. The city has already opened a second line, with plans for three more in the wake of its success.

Light rail, or trolleys, are also enjoying a global revival. Old systems are being renovated and many cities, even in the car-dominated US, are opening new lines. In the US, a new type of housing, "transit-oriented development" (high-density development centered around a transit stop), is proving popular as it allows people to commute easily to work, without having to drive.

Rail travel has succeeded by seeing off the alternatives. For a time other technologies were variously put forward as having greater potential to improve transportation, including numerous bizarre monorail plans. The most prominent of these was "maglev"—magnetic levitation (see pp.388–89). Magnetic force is used to elevate the "train" slightly above the special track and then magnets are used to provide forward thrust. The result was a very smooth ride at far higher speeds than conventional trains, along with better acceleration and braking. Despite many decades of research and development, and the introduction of a few systems, there are still currently only two maglev systems in operation—one in Japan and the other in China. The Chinese maglev, shuttling passengers between Shanghai and the airport, takes just seven minutes 20 seconds to cover 18¾ miles (30km) and reaches a speed of 268mph (430kph). However, the cost of development, the potential risks (there has been one major accident on a German test track, resulting in 23 deaths), and the fact that conventional rail is already a tried and tested technology used across the world, has meant that maglev expansion has been stymied. While there are proposals for lines in several countries, it is clear there is no immediate prospect of this technology replacing rail.

In what has proved to be a real shock to many transportation planners (and past futurologists), the railroads have not only survived to see the 21st century, but are, in fact, booming. As oil becomes scarcer and concerns about environmental impact grow, rail travel will only appear more attractive. Rail offers convenience, safety, and speed, as well as compatability with personal technology: travelers can use mobile devices or work on laptops, using time otherwise wasted behind the wheel. The train is becoming more, not less, suited to life in today's world. The 21st century will be the second age of the train.

Maglev Trains

Unlike conventional trains with their track-and-wheel interface, maglev (magnetic levitation) trains literally float on air, using powerful magnets to suspend trains at a constant level above a steel rail or guideway and electromagnetic force to propel the trains. Due to a lack of friction, maglev trains are quiet and stable, can accelerate and decelerate fast, and both trains and guideway suffer little wear and tear. However, due to incompatibility with existing railroads, maglev technology has seen limited adoption—the only working commercial systems are in Japan and China. The maglev infrastructure is expensive to build, but once in place operating costs are low and the trains can achieve very high velocity: maglev trains hold the world speed record for rail transportation.

Alternative Systems

The SCMaglev (Superconducting Maglev, named for the train's powerful magnets) is the latest in a series of high-speed maglev trains developed in Japan. It makes use of electrodynamic suspension (EDS) on U-shaped rails for levitation and propulsion. It has undergone successful trials, though has not yet progressed to commercial use.

THE SCMAGLEV IS THE FASTEST PASSENGER TRANSPORT TRAIN IN THE WORLD, REACHING 361MPH (581KPH) ON A TEST TRACK IN 2003.

Train runs on a T-shaped monorail which makes derailment almost impossible

How it works

Existing commercial maglev systems use electromagnetic suspension (EMS), in which magnets in the train are activated for both levitation and propulsion by an electric current in the track. This current can be adjusted to determine the train's speed, while electronic sensors monitor the gap between train and guideway.

LEVITATION
Powerful electromagnets on the undercarriage of the train, which wraps around the T-shaped guideway, are attracted by levitation and guidance coils mounted in the rail.

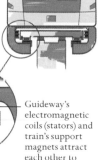

Guide magnets on train maintain an ⅜–½in (8–12mm) gap

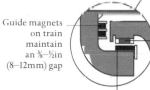

Guideway's electromagnetic coils (stators) and train's support magnets attract each other to levitate train

Battery-powered support magnets on train

PROPULSION
The polarity of propulsion coils in the rail changes constantly, attracting and repelling magnets on the train. The frequency is reversed to stop the train.

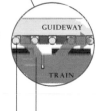

GUIDEWAY

TRAIN

Guideway's alternating current attracts then repels train's magnets to propel train

Direction of magnetic field, reversed to brake

TRANSRAPID MAGLEV
German company, Transrapid, the leader in maglev trains, has been honing its electromagnetic suspension (EMS) since the 1960s. Seen here on its test track, the Transrapid train went into commercial operation in 2004 on China's Shanghai Maglev, which travels the 18½ miles (30km) from the airport to the financial district in 8 minutes at 250mph (400kph).

Lack of engine makes maglev trains light and energy-efficient: air cooling the propulsion system uses more power than levitation

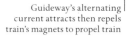

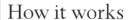

C-shaped undercarriage contains the train's support magnets for levitation

Glossary

ADHESION The frictional grip between the wheel of a train and the rail of a track.

AIR CUSHION A "spring" of air used in modern suspension systems.

AIR BRAKE A braking system that uses compressed air as its operating medium.

AMERICAN A steam locomotive with a *wheel arrangement* of 4-4-0.

ATLANTIC A steam locomotive with a *wheel arrangement* of 4-4-2.

BALDWIN A US *locomotive* manufacturer that was in business from 1825 to 1971.

BALLAST The bed of stone, gravel, or cinders on which a rail track is laid.

BANK A steep section of a track that a train requires additional engines to climb.

BERKSHIRE A steam *locomotive* with a *wheel arrangement* of 2-8-4.

BLASTPIPE The exhaust pipe of a steam *locomotive* that diverts steam from the cylinders into the *smokebox* beneath the *smokestack* to increase the draft through the fire.

BOGIE See *truck*

BOILER Cylindrical chamber in which steam is produced to drive a steam *locomotive*.

BOXCAR An enclosed rectangular freight car that has doors and is used for general service and especially for lading that must be protected from the weather. Called a *van* in the UK.

BRANCH LINE A secondary railroad line that branches off a *main line*.

BROAD GAUGE Rails spaced more widely than the *standard gauge* of 4ft 8½ in (1,435mm).

BUMPING POST / BUMPER The post at the end of a track that halts a train from traveling any further.

CAB The control room of a *locomotive*, providing shelter and seats for the engine crew.

CABOOSE A car attached to the end of a freight train and used as an office and headquarters for the conductor and brakeman while in transit.

CANT The difference of elevation of a rail relative to its partner rail.

CHALLENGER A steam *locomotive* with a *wheel arrangement* of 4-6-6-4.

CHIMNEY See *smokestack*

COMPOUND LOCOMOTIVE A steam *locomotive* that uses two sets of cylinders, the second powered by exhaust steam from the first.

COUPLER A device located at both ends of rail cars in a standard location to provide a means for connecting one rail car to another. Called a *coupling* in the UK.

COUPLING ROD A rod that connects driving wheels on a steam locomotive.

COWCATCHER A metal frame projecting from the front of a steam *locomotive* designed to clear the track of obstructions.

CUTTING / CUT A channel dug through the hillside to enable a rail track to maintain a shallow gradient.

CYLINDER The core of a steam engine in which a *piston* moves back and forth under the pressure of expanding and condensing steam.

DRAFT GEAR A term used to describe the shock absorbing unit that forms the connection between the coupler and the center sill.

EMBANKMENT A raised pathway across a depression in the landscape that enables a rail track to maintain a shallow gradient.

ENGINE The power source of a *locomotive*, driven by steam, electricity, or diesel.

EXPRESS TRAIN A train that passes certain stations on its route without stopping, to arrive at its final destination faster, as opposed to a local train, which makes all stops along its route.

FIREBOX The compartment within a steam engine where fuel is burned to provide heat.

FIREMAN / STOKER / BOILERMAN A worker responsible for keeping the *firebox* of an *engine* fed with coal.

FREIGHT / GOODS Materials or products transported for commercial gain.

GANDY DANCER An early term for a track maintenance worker.

GAUGE The width between the inner faces of the rails.

GONDOLA An open-top piece of *rolling stock* that has straight sides and ends and a level floor; used for bulk freight, such as steel pipes and rolls of cable.

GRADE CROSSING A location where a railroad crosses a road or path at the same elevation.

HANDCAR A small, open railroad car propelled by its passengers, often by means of a hand pump. Known as a *pump trolley* in Britain.

HUDSON A steam *locomotive* with a *wheel arrangement* of 4-6-4.

INTERCHANGE The transfer of cars from one railroad to another at a common junction point.

INTERLOCKING TOWER A control room in which the movement of trains is controlled by means of signals and blocks, ensuring trains travel safely and on schedule. The UK term is signal box.

INTERMODAL A flexible way of transporting freight over water, highway, and rail without it being removed from the original transportation equipment, namely a container trailer.

JUBILEE A steam locomotive with a *wheel arrangement* of 4-6-4.

JUNCTION A place where multiple train lines split or converge.

LANTERN A portable lamp with a fuel source. Used by early railroad workers to provide light and to signal to other workers at night.

LIGHT RAIL Small, fixed railroads, typically operating within urban environments e.g. streetcars and trolleys.

LOCOMOTIVE The engine-powered vehicle that either pulls or pushes a train along the tracks.

LOOP A railroad formation where tracks cross over themselves as they ascend a mountain.

MAIN LINE The primary line between major towns or cities, exclusive of branch lines.

MARCHALING YARD See *yard*

MIKADO A steam *locomotive* with the *wheel arrangement* 2-8-2.

MONORAIL A railroad system based on a single rail. Often elevated, and built in urban environments.

NARROW GAUGE A railroad with a gauge narrower than the standard 4ft 8½ in (1,435mm).

NAVVIES A British term for specialized manual laborers who constructed the majority of the railroads in the 19th century.

PACIFIC A steam *locomotive* with a *wheel arrangement* of 4-6-2.

PANTOGRAPH A metal arm that slides underneath an overhad electric line, providing power to an electric train.

PASSENGER TRAIN A train with passenger cars intended to transport people.

PASSING LOOP See *siding*

PASSING SIDING A position on a single-track railroad, where trains traveling in opposite directions can pass each other. Called a *passing loop* in the UK.

PISTON A component of an internal combustion engine which moves up and down against a liquid or gas to provide motion.

POINTS / RAILROAD SWITCH A movable section of a railroad that allows a train to move from one track to another.

PRAIRIE A steam locomotive with a *wheel arrangement* of 2-6-2.

RAILROAD CAR A covered railroad vehicle used for carrying passengers or cargo.

ROLLING STOCK Used by railroad companies to refer to the entire collection of equipment that runs run on their railroad.

ROUNDHOUSE Buildings used to service and store locomotives. Many used to be arranged around a *turntable.*

SIDING A section of track off the main line used for passing trains or for storing *rolling stock.*

SIGNAL BOX See *interlocking tower*

SLEEPER A train that can provide beds for its passengers, particularly for overnight or long-distance journeys.

SMOKEBOX A component of a steam engine. The smokebox would collect smoke from the *firebox*, after it had heated water to provide steam, and release it though the *smokestack* or *chimney.*

SMOKESTACK The vertical exhaust funnel of a train. Called a *chimney* in the UK.

STANDARD GAUGE Rails spaced 4ft 8½ in (1,435mm) apart.

STATION MASTER The individual in charge of running a station.

STEAM ENGINE An engine that uses steam, produced by heating water with burning fuel, to perform mechanical work.

SUBWAY A railroad that operates primarily underground, typically in a major city. Known as the *underground* in the UK.

TENDER The car attached immediately behind a steam locomotive containing the necessary fuel and water needed to power a steam locomotive.

THROUGH COACH A passenger car that switches *locomotives* mid-journey, removing the need for passengers to switch trains. Used particularly on long haul journeys.

TRACK The permanent fixtures of rails, *ballast, fastenings*, and underlying substrate that provide a runway for the wheels of a train.

TRACTION The act of drawing or pulling a load. Can also refer to the adhesive friction of a train to a track.

TRUCK The undercarriage assembly of a train, incorporating the wheels, suspension, and brakes. Called a *bogie* in the UK.

TURNTABLE A device for rotating rail vehicles so they can travel back in the direction they came from. Largely obsolete today.

UNDERGROUND See *subway*

UNIT TRAIN A train that carries only one type of good or commodity, e.g., coal.

VAN See *boxcar*

WATER CRANE See *water tower*

WATER TOWER A track-side device for quickly refilling the water tank of a steam *locomotive.* Known in the UK as a *water crane.*

WHEEL The wheels of trains are typically cast or forged from hardened steel. .

WHEEL ARRANGEMENT A system for classifying how wheels are placed under a *locomotive*, such as the *Whyte notation.*

WHEEL FLANGE A component of a train wheel. The flange extends the wheels to the interior of the train track, preventing the train from running off the rails.

WHYTE NOTATION A system classifying *wheel arrangment* by counting first leading wheels, then driving wheels, then trailing wheels (e.g. 0-2-2).

YARD An area with multiple tracks, other than main tracks, and sidings for the storage, maintenance, and loading and unloading of *rolling stock*, and where freight cars are organized into trains.

YELLOWSTONE A steam *locomotive* with a *wheel arrangement* of 2-8-8-4.

ZIGZAG / SWITCHBACK A method of track construction on steep inclines. A train ascends and descends the track in a zigzag fashion.

Bibliography

This is very much a selective bibliography, mainly mentioning books I have used as source material, since there are literally tens of thousands of books on the railroads. Many of these are very detailed and written for a specialist audience, and have consequently not been included in this list. The list is, therefore, aimed at the general reader who wants to know more on the subjects covered in this book, rather than at a specialist audience.

I have, of course, made extensive use of my own series of six railroad history books, all published by Atlantic. *The Subterranean Railway* (2004, updated 2013) is the story of the London Underground, *Fire and Steam* (2006) covers the story of Britain's railroads, and *Blood, Iron and Gold* (2008) shows how the railroads changed the world. *Engines of War* (2010) demonstrates the importance of railroads in wartime while *The Great Railway Revolution* (2012) is the story of American railroads; *To the Edge of the World* (2013) is the history of the world's longest railroad, the Trans-Siberian.

GENERAL

Erwin Berghaus, *The History of the Railways*, Barrie & Rockliffe, 1964

Anthony Burton, *Railway Empire*, John Murray, 1994

Anthony Burton, *On the Rails*, Aurum, 2004

Christopher Chant, *The World's Railways*, Grange, 2002

Basil Cooper, *A Century of Train*, Brian Trodd Publishing, 1988

Nicholas Faith, *Locomotion*, BBC Books, 1993

Nicholas Faith, *The World the Railways Made*, Bodley Head, 1990

Tim Fischer, *Trains Unlimited*, ABC Books, 2011

Geoffrey Freeman Allen, *Railways Past, Present and Future*, Orbis Publishing, 1982

Geoffrey Freeman Allen, *Railways of the Twentieth Century*, Winchmore, 1983

Geoffrey Freeman Allen, *Luxury Trains of the World*, Bison, 1979

Jim Harter, *World Railways of the Nineteenth Century: A Pictorial History in Victorian Engravings*, Johns Hopkins University Press, 2005

Clive Lamming, *Larousse des Trains et des Chemins de Fer*, Larousse, 2005

Bryan Morgan, ed, *Great Trains*, Crown Publishers, 1973

O.S. Nock, *World Atlas of Railways*, Mitchell Beazley, 1978

O.S. Nock, *Railways Then and Now: A World History*, Paul Elek Ltd, 1975

O.S. Nock, ed, *Encyclopaedia of Railways*, Book Club Associates, 1977

Martin Page, *The Lost Pleasures of the Great Trains*, Weidenfeld and Nicolson, 1975

Steve Parissien, *Station to Station*, Phaidon, 1997

P.J.G. Ransom, *Locomotion: Two Centuries of Train Travel*, Sutton Publishing, 2001

Michael Robbins, *The Railway Age*, Penguin, 1965

Wolfgang Schivelbusch, *Railway Journey: The Industrialization of Time and Space in the Nineteenth Century*, Berg, 1996

John Westwood, *Railways at War*, Osprey, 1980

John Westwood, *The Pictorial History of Railways*, Bison Books, 2008

EUROPE

H.C. Casserly, *Outline of Irish History*, David & Charles, 1974

Nicholas Faith, *The Right Line: the Politics, the Planning and the Against-the-odds Gamble Behind Britain's First High-speed Railway*, Segrave Foulkes, 2007

Peter Fleming, *The Fate of Admiral Kolchak*, Rupert Hart David, 1963 (reprinted 2001 by Birlinn)

Murray Hughes, *Rail 300*, David & Charles, 1988

P.M. Kalla-Bishop, *Italian Railroads*, Drake, 1972

P.M. Kalla-Bishop, *Mediterranean Island Railways*, David & Charles, 1970

Allan Mitchell, *The Great Train Race: Railways and Franco-German Rivalry*, Berghahn, 2000

O.S. Nock, *Railways of Western Europe*, A&C Black, 1977

Brian Perren, *TGV Handbook, Capital Transport*, 1998

Albert Schram, *Railways and the Formation of the Italian State in the Nineteenth Century*, Cambridge University Press, 1977

Christine Sutherland, *The Princess of Siberia*, Methuen, 1984

Various authors, *Histoire du Réseau Ferroviaire Français*, Editions de l'Ormet, 1996

Various authors, *ICE: High-Tech on Wheels*, Hestra-Verlag, 1991

Arthur J. Veenendaal, *Railways in the Netherlands: A Brief History, 1834–1994*, Stanford University Press, 2001

THE AMERICAS

Dee Brown, *Hear That Lonesome Whistle Blow: Railroads in the West*, Touchstone, 1977

David Cruise and Alison Griffiths, *Lords of the Line: The Men Who Built the Canadian Pacific Railway*, Viking, 1988

Brian Fawcett, *Railways of the Andes*, Plateway Press, 1997

Sarah H. Gordon. *Passage to Union: How the Railroads Transformed American Life, 1829–1929*, Elephant Paperbacks, 1997

George W. Hilton and John F. Due, *The Electric Interurban Railways in America*, Stanford University Press, 1960

Stewart H. Holbrook, *The Story of American Railroads*, Bonanza Books, 1947

Theodore Kornweibel Jr, *Railroads in the African American Experience*, Johns Hopkins University Press, 2010

Oscar Lewis, *The Big Four*, Alfred A. Knopf, 1938

Albro Martin, *Railroads Triumphant*, Oxford University Press, 1992

Nick and Helma Mika, *The Railways of Canada: A Pictorial History*, McGraw-Hill Ryerson, 1972

O.S. Nock, *Railways of Canada*, A&C Black, 1973

Andrew Roden, *Great Western Railway: A History*, Aurum, 2010

David Rollinson. *Railways of the Caribbean*, Macmillan, 2001

D. Trevor Rowe, *The Railways of South America*, Locomotives International, 2000

John F. Stover, *American Railroads*, University of Chicago Press, 1961

Richard White, *The Transcontinentals and the Making of Modern America*, Norton, 2011

Oscar Zanetti and Alejandra García, *Sugar and Railroads: A Cuban History, 1837–1959*, University of North Carolina Press, 1998

ASIA

Ralph William Huenemann, *The Dragon and the Iron Horse: The Economics of Railroads in China, 1876–1937*, Harvard University Press, 1984

Robert Hardie, *The Burma Siam Railway*, Quadrant Books, 1984

Ian J. Kerr, Engines of Change: *The Railways that Made India*, Praeger, 2007

Ian J. Kerr, *Building the Railways of the Raj, 1850–1900*, Oxford University Press, 1995

Abrahm Lustgarten, *China's Great Train: Beijing's Drive West and the Campaign to Remake Tibet*, Henry Holt, 2008

Deborah Manley, ed, *The Trans-Siberian Railway: A Traveller's Anthology*, Century Hutchinson, 1987

Steven G. Marks, *Road to Power: The Trans-Siberian Railroad and the Colonization of Asian Russia, 1850–1917*, Cornell University Press, 1991

James Nicholson, *The Hejaz Railway*, Stacey International, 2005

O.S. Nock, *Railways of Asia and the Far East*, A&C Black, 1978

Peter Semmens, *High Speed in Japan*, Platform 5, 2000

Roopa Srinivasan, Manish Tiwari, and Sandeep Silas, *Our Indian Railway*, Foundation Books, 2006

Shoji Sumita *Success Story, the privatisation of Japanese National Railways*, Profile Books, 2000

John Tickner, Gordon Edgar, and Adrian Freeman, *China: The World's Last Steam Railway*, Artists' and Photographers' Press, 2008

Harmon Tupper, *To the Great Ocean*, Secker & Warburg, 1965

K.R. Vaidyanathan, *150 Glorious Years of Indian Railways*, English Edition Publishers, 2003

Christopher J. Ward, *Brezhnev's Folly: The Building of the BAM and Late Soviet Socialism*, University of Pittsburgh Press, 2009

Various authors, *Guide to the Great Siberian Railway*, 1900, David & Charles reprints, 1971

AFRICA

John Day, *Railways of South Africa*, Arthur Barker, 1963

M.F. Hill, *The Permanent Way: The Story of the Tanganyika Railways*, East African Railways and Harbours, 1958

George Tabor, *Cape to Cairo*, Genta, 2003

AUSTRALASIA

Neill Atkinson, *Trainland*, Random House, 2007

Tim Fischer, *Transcontinental Train Journey*, Allen & Unwin, 2004

C.C. Singleton and David Burke, *Railways of Australia*, Angus & Robertson, 1963

Patsy Adam Smith, *The Desert Railway*, Rigby, 1974

Patsy Adam Smith, *Romance of Australian Railways*, Rigby, 1973

Index

Illustrations are in *italics*

Acknowledgments

The author would like to thank Nicholas Faith, author of *The World the Railways Made*, for drafting several chapters and advising on various aspects of the book, and Malcolm Bulpitt of the Swiss Railway Society for his draft of the section on Switzerland.

The publisher would like to thank the following for their kind permission to reproduce their photographs:

(Key: a-above; b-below/bottom; c-centre; f-far; l-left; r-right; t-top)

2 Matthew Malkiewicz: losttracksoftime.com. 5 Matthew Malkiewicz: losttracksoftime.com. 11 Science & Society Picture Library: NRM / Pictorial Collection. 15 Getty Images: Lonely Planet Images. 18-19 Science & Society Picture Library: Science Museum. 20-21 Science & Society Picture Library: Science Museum. 22 The Bridgeman Art Library: Institute of Mechanical Engineers, London, UK. 23 DK Images: Courtesy of the National Railway Museum, York. 25 Science & Society Picture Library: NRM. 26-27 Science & Society Picture Library: NRM / Pictorial Collection. 28 Science & Society Picture Library: Science Museum. 30-31 Science & Society Picture Library: NRM. 32 Smithsonian Institution Archives: NPG.75.13. 35 DK Images: Courtesy of Railroad Museum of Pennsylvania. 36 Corbis: Bettmann. 37 38 DK Images: Courtesy of B&O Railroad Museum. 40 DK Images: Courtesy of B&O Railroad Museum (cl, cr, bl). 41 DK Images: Courtesy of B&O Railroad Museum (tl). 41 DK Images: Courtesy of Railroad Museum of Pennsylvania (tr, cl, cr). 40-41b DK Images: Courtesy of B&O Railroad Museum. 42 Getty Images: Culture Club. 45 Topfoto: ullsteinbild. 47 Alamy: The Art Archive. 48 Getty Images. 53 Corbis: Michael Nicholson. 54 The Bridgeman Art Library: Sunderland Museums & Winter Garden Collection, Tyne & Wear, UK. 57 Science & Society Picture Library: NRM / Cuneo Fine Arts (Artist copyright Estate of Terence Cuneo / The Bridgeman Art Library). 58 Traditionsbetriebswerk Stassfurt: (bl). 58-59 Getty Images: Fox Photos. 66-67 Corbis: Horace Bristol. 61 AF Eisenbahn Archiv. 62 Library of Congress, Washington, D.C.. 65 The Kobal Collection: United Artists. 66 Raggan Datta: (bc). 68 DK Images: Courtesy of Didcot Railway Centre. 72-73 Science & Society Picture Library: NRM / Pictorial Collection. 74-75 akg-images. 77 Getty Images: Hulton Archive. 79 Science & Society Picture Library: Science Museum. 80 Getty Images: Hulton Archive. 82 Getty Images: The British Library / Robana (bl). 85 Science & Society Picture Library: NRM. 86 Science & Society Picture Library: Science Museum. 89 Getty Images: Sovfoto / UIG. 90-91 Science & Society Picture Library: NRM. 92 Mary Evans: Iberfoto. 93 Fotolia.com: cityanimal. 96-97 Getty Images: The British Library / Robana. 98 Alamy: 19th era. 99 AF Eisenbahn Archiv. 102 AF Eisenbahn Archiv. 104 Alamy: imagebroker. 107 Corbis: Swim Ink 2, LLC. 108-109 Getty Images: Roy Stevens / Time & Life Pictures. 110 Topfoto: The Granger Collection. 113 William L. Clements Library, University of Michigan. 115 California State Library. 116 Mary Evans: Everett Collection. 118-119 Corbis: Michael Maslan Historic Photographs. 120 California State Library. 123 Corbis: Bettmann. 124 Corbis: James L. Amos. 126 Canadian Pacific Railway. 128 The Bridgeman Art Library: Private Collection / Peter Newark American Pictures (tl). 132-133 Science & Society Picture Library: Science Museum. 134 Getty Images: Hulton Archive (b). 137 Getty Images: Keystone / Hulton Archive (tr). 139 National Museums Northern Ireland: Collection Armagh County Museum (b). 140 Getty Images: Jimin Lai / AFP (b). 142 Luped.com: Roland Smithies (t). 144 DK Images: Courtesy of Railroad Museum of Pennsylvania (bl). 144-145 akg-images: North Wind Picture Archives. 147 Luped.com: Roland Smithies. 148 SuperStock: Christie's Images Ltd. 150 DK Images: Courtesy of B&O Railroad Museum. 151 Getty Images: MPI. 152 Corbis: Underwood & Underwood (br). 152 Getty Images: Digital Vision (bl). 153 Science & Society Picture Library: NRM / Cuneo Fine Arts (bl). 156 Getty Images: Popperfoto. 159 Getty Images. Hal Morey. 160 Science & Society Picture Library. Science Museum (t). 161 DK Images: Courtesy of Railroad Museum of Pennsylvania (t). 161 Getty Images: Fox Photos (b). 163 Corbis: Underwood & Underwood. 165 AF Eisenbahn Archiv. 165 Getty Images: Fotosearch. 166 Library of Congress, Washington, D.C.. 168 Alexander Turnbull Library, Wellington, New Zealand: EP-Accidents-Rail-Tangiwai rail disaster-01 (bl). 168-169 Science & Society Picture Library: NRM. 170 Getty Images: Chicago History Museum. 172 Corbis: Bettmann. 174-175 AF Eisenbahn Archiv. 177 Getty Images: MPI. 181 The Bridgeman Art Library: Regional Art Museum, Irkutsk. 183 AF Eisenbahn Archiv. 184 Getty Images: Sovfoto / UIG. 187 Getty Images: De Agostini / E. Ganzerla. 188 Corbis: Hulton-Deutsch Collection (bl). 190 Mary Evans: Epic. 192 Getty Images: Culture Club. 193 The Art Archive: Kharbine-Tapabor / Collection IM. 194-195 Corbis: Katie Garrod / JAI. 196 akg-images. 198 South American Pictures. 201 Topfoto: Alinari. 202 Wikimedia: Ernesto Linares. 204 Alamy: Prisma Bildagentur AG (bl). 205 Getty Images: Andrey Rudakov / Bloomberg (cl). 204-205 Marcelo Meneses / Ecuador Adventure: (b). 208 AF Eisenbahn Archiv. 210-211 The Art Archive: Culver Pictures. 212 Alamy: imagebroker (cla). 212 DK Images: Courtesy of B&O Railroad Museum (b). 213 Alamy: Gerry White (cla). 213 Alamy: David

Acknowledgments

The author would like to thank Nicholas Faith, author of *The World the Railways Made*, for drafting several chapters and advising on various aspects of the book, and Malcolm Bulpitt of the Swiss Railway Society for his draft of the section on Switzerland.

The publisher would like to thank the following for their kind permission to reproduce their photographs:

(Key: a-above; b-below/bottom; c-centre; f-far; l-left; r-right; t-top)

2 Matthew Malkiewicz: losttracksoftime.com. 5 Matthew Malkiewicz: losttracksoftime.com. 11 Science & Society Picture Library: NRM / Pictorial Collection. 15 Getty Images: Lonely Planet Images. 18-19 Science & Society Picture Library: Science Museum. 20-21 Science & Society Picture Library: Science Museum. 22 The Bridgeman Art Library: Institute of Mechanical Engineers, London, UK. 23 DK Images: Courtesy of the National Railway Museum, York. 25 Science & Society Picture Library: NRM. 26-27 Science & Society Picture Library: NRM / Pictorial Collection. 28 Science & Society Picture Library: Science Museum. 30-31 Science & Society Picture Library: NRM. 32 Smithsonian Institution Archives: NPG.75.13. 35 DK Images: Courtesy of Railroad Museum of Pennsylvania. 36 Corbis: Bettmann. 37-38 DK Images: Courtesy of B&O Railroad Museum. 40 DK Images: Courtesy of B&O Railroad Museum (cl, cr, bl). 41 DK Images: Courtesy of B&O Railroad Museum (tl). 41 DK Images: Courtesy of Railroad Museum of Pennsylvania (tr, cl, cr). 40-41b DK Images: Courtesy of B&O Railroad Museum. 42 Getty Images: Culture Club. 45 Topfoto: ullsteinbild. 47 Alamy: The Art Archive. 48 Getty Images. 53 Corbis: Michael Nicholson. 54 The Bridgeman Art Library: Sunderland Museums & Winter Garden Collection, Tyne & Wear, UK. 57 Science & Society Picture Library: NRM / Cuneo Fine Arts (Artist copyright Estate of Terence Cuneo / The Bridgeman Art Library). 58 Traditionsbetriebswerk Stassfurt: (bl). 58-59 Getty Images: Fox Photos. 66-67 Corbis: Horace Bristol. 61 AF Eisenbahn Archiv. 62 Library of Congress, Washington, D.C.. 65 The Kobal Collection: United Artists. 66 Raggan Datta: (bc). 68 DK Images: Courtesy of Didcot Railway Centre. 72-73 Science & Society Picture Library: NRM / Pictorial Collection. 74-75 akg-images. 77 Getty Images: Hulton Archive. 79 Science & Society Picture Library: Science Museum. 80 Getty Images: Hulton Archive. 82 Getty Images: The British Library / Robana (bl). 85 Science & Society Picture Library: NRM. 86 Science & Society Picture Library: Science Museum. 89 Getty Images: Sovfoto / UIG. 90-91 Science & Society

Picture Library: NRM. 92 Mary Evans: Iberfoto. 93 Fotolia.com: cityanimal. 96-97 Getty Images: The British Library / Robana. 98 Alamy: 19th era. 99 AF Eisenbahn Archiv. 102 AF Eisenbahn Archiv. 104 Alamy: imagebroker. 107 Corbis: Swim Ink 2, LLC. 108-109 Getty Images: Roy Stevens / Time & Life Pictures. 110 Topfoto: The Granger Collection. 113 William L. Clements Library, University of Michigan. 115 California State Library. 116 Mary Evans: Everett Collection. 118-119 Corbis: Michael Maslan Historic Photographs. 120 California State Library. 124 Corbis: Bettmann. 126 Canadian Pacific Railway. 128 The Bridgeman Art Library: Private Collection / Peter Newark American Pictures (tl). 132-133 Science & Society Picture Library: Science Museum. 134 Getty Images: Hulton Archive (b). 137 Getty Images: Keystone / Hulton Archive (tr). 139 National Museums Northern Ireland: Collection Armagh County Museum (b). 140 Getty Images: Jimin Lai / AFP (b). 142 Luped.com: Roland Smithies (t). 144 DK Images: Courtesy of Railroad Museum of Pennsylvania (bl). 144-145 akg-images: North Wind Picture Archives. 147 Luped.com: Roland Smithies. 148 SuperStock: Christie's Images Ltd. 150 DK Images: Courtesy of B&O Railroad Museum. 151 Getty Images. MPI. 152 Corbis. Underwood & Underwood (br). 152 Getty Images: Digital Vision (bl). 153 Science & Society Picture Library: NRM / Cuneo Fine Arts (bl). 156 Getty Images: Popperfoto. 159 Getty Images. Hal Morey. 160 Science & Society Picture Library: Science Museum (c). 161 DK Images: Courtesy of Railroad Museum of Pennsylvania (t). 161 Getty Images: Fox Photos (b). 163 Corbis: Underwood & Underwood. 165 AF Eisenbahn Archiv. 165 Getty Images: Fotosearch. 166 Library of Congress, Washington, D.C.. 168 Alexander Turnbull Library, Wellington, New Zealand: EP-Accidents-Rail-Tangiwai rail disaster-01 (bl). 168-169 Science & Society Picture Library: NRM. 170 Getty Images: Chicago History Museum. 172 Corbis: Bettmann. 174-175 AF Eisenbahn Archiv. 177 Getty Images: MPI. 181 The Bridgeman Art Library: Regional Art Museum, Irkutsk. 183 AF Eisenbahn Archiv. 184 Getty Images: Sovfoto / UIG. 187 Getty Images: De Agostini / E. Ganzerla. 188 Corbis: Hulton-Deutsch Collection (bl). 190 Mary Evans: Epic. 192 Getty Images: Culture Club. 193 The Art Archive: Kharbine-Tapabor / Collection IM. 194-195 Corbis: Katie Garrod / JAI. 196 akg-images. 198 South American Pictures. 201 Topfoto: Alinari. 202 Wikimedia: Ernesto Linares. 204 Alamy: Prisma Bildagentur AG (bl). 205 Getty Images: Andrey Rudakov / Bloomberg (cl). 204-205 Marcelo Meneses / Ecuador Adventure: (b). 208 AF Eisenbahn Archiv. 210-211 The Art Archive: Culver Pictures. 212 Alamy: imagebroker (cla). 212 DK Images: Courtesy of B&O Railroad Museum (b). 213 Alamy: Gerry White (cla). 213 Alamy: David

Davies (br). **213 DK Images:** Courtesy of Virginia Museum of Transport (tl, cra). **212-213c DK Images:** Courtesy of B&O Railroad Museum. **214 Topfoto:** The Granger Collection. **215 Fotolia.com:** Popova Olga. **216 Mary Evans Picture Library. 219 The Art Archive:** Eileen Tweedy. **221 Corbis:** Ocean. **222 Mary Evans:** Illustrated London News Ltd (bl). **224-225 Getty Images:** General Photographic Agency. **226 Science & Society Picture Library:** Science Museum. **228 Topfoto:** The Granger Collection. **230 DK Images:** Courtesy of Railroad Museum of Pennsylvania (cl). **230 DK Images:** Courtesy of Ribble Stram Railway (cra). **230 DK Images:** Courtesy of the National Railway Museum, York (cla). **231 Alamy:** pf (tr). **231 DK Images:** Courtesy of Railroad Museum of Pennsylvania (cl). **231 DK Images:** Courtesy of the Musee de Chemin de Fer, Mulhouse (tl). **231 Getty Images:** Tomohiro Ohsumi / Bloomberg (cr). **230-231b DK Images:** Courtesy of Railroad Museum of Pennsylvania. **233 Corbis:** S W A Newton / English Heritage / Arcaid. **234-235 The Bridgeman Art Library:** Hagley Museum & Library, Wilmington, Delaware, USA. **236 Bonhams. 238 Science & Society Picture Library:** NRM / Pictorial Collection. **241 Topfoto. 242 Corbis:** Bettmann. **246-247 Getty Images:** Stock Montage. **249 Getty Images:** Hulton Archive. **250-251 AF Eisenbahn Archiv. 254 AF Eisenbahn Archiv. 252-253 DK Images:** Courtesy of Adrian Shooter. **256 Mary Evans Picture Library. 261 Mary Evans:** Grenville Collins Postcard Collection. **262 Topfoto:** The Granger Collection. **265 Corbis:** Scheufler Collection. **266 Corbis. 268-269 Getty Images:** Fox Photos. **271 Alamy:** The Print Collector. **272 Topfoto. 275 The Bridgeman Art Library:** Galerie Bilderwelt. **276 DK Images:** Courtesy of Virginia Museum of Transportation (b). **276 DK Images:** Courtesy of Railroad Museum of Pennsylvania (tr). **276 DK Images:** Courtesy of Railroad Museum of Pennsylvania (ca). **277 DK Images:** Courtesy of B&O Railroad Museum (t, c, b). **276-277 DK Images:** Courtesy of Virginia Museum of Transportation (c). **278-279 Getty Images:** Neurdein / Roger Viollet. **280-281 AF Eisenbahn Archiv. 284 REX Features:** Sipa Press. **288 Getty Images:** Hulton Archive. **291 Corbis:** Col. F. R. Maunsell / National Geographic Society. **293 Corbis:** John Springer Collection. **294 DK Images:** Courtesy of Railroad Museum of Pennsylvania (bl). **295 DK Images:** Courtesy of Virginia Museum of Transportation (t). **295 DK Images:** Courtesy of the DB Museum Nürnberg, Germany (cb). **295 Getty Images:** The Asahi Shimbun (br). **294-295 DK Images:** Courtesy of the National Railway Museum, York (c). **298 Alamy:** imagebroker. **300-301 Alamy:** WoodyStock. **302 Alexander Turnbull Library, Wellington, New Zealand:** Ref: Eph-E-RAIL-1940s-01. **303 N Z Railway & Locomotive Society Collection:** J D Buckley. **305 Corbis:** Hulton-Deutsch Collection. **306 Science & Society Picture Library:** NRM / Pictorial Collection. **309 DK Images:** Courtesy of Railroad Museum of Pennsylvania. **310-311 Science & Society Picture Library:** NRM / Pictorial Collection (Artist copyright Gerald Coulson). **313 Topfoto:** ullsteinbild. **315 Corbis:** Underwood & Underwood. **316 AF Eisenbahn Archiv. 318 Alamy:** Daniel Dempster Photography (br). **318 Alamy:** John Gaffen 2 (bl). **318 DK Images:** Courtesy of Railroad Museum of Pennsylvania (c). **319 Alamy:** Thomas J. Peterson (tl). **319 Alamy:** Frank Paul (b). **319 DK Images:** Courtesy of the Museum of Transportation, St. Louis, Missouri (tr). **319 Science & Society Picture Library:** NRM (cr). **321 Corbis:** Bettmann. **323 Australian War Memorial:** Order 6459212. **324 Getty Images:** LatitudeStock / Emma Durnford. **326-327 SuperStock:** Andrew Woodley / age fotostock. **332 Getty Images:** Nadezhda Borovaya Estate / Chip HIRES / Gamma-Rapho. **334 Luped. com. 335 Alamy:** The Print Collector. **337 Alamy:** RIA Novosti. **337-338 Getty Images:** John Mueller. **343 Alamy:** Keystone Pictures USA. **344 Alamy:** Inge Johnsson. **346-347 Matthew Malkiewicz:** losttracksoftime.com. **348 Getty Images:** DeAgostini. **350-351 Alamy:** qaphotos.com. **353 Getty Images:** Denis Charlet / AFP. **354 Getty Images:** James L. Stanfield / National Geographic (bl). **354-355 Alamy:** qaphotos.com. **357 Corbis:** Hulton-Deutsch Collection (b). **358 Corbis:** Swim Ink 2, LLC (t). **359 Corbis:** Daniel Schoenen / imagebroker. **362-363 Alamy:** age fotostock. **365 Getty Images:** Sankei Archive. **366-367 Dreamstime.com:** Sean Pavone. **371 Alamy:** Jon Arnold Images Ltd. **372 SuperStock:** imagebroker. net. **375 Corbis:** Wu Hong / epa. **376 DK Images:** Courtesy of the National Railway Museum, York. **379 Reuters:** Darley Shen. **380-381 Corbis:** Chen Xie / Xinhua Press. **382-383 age fotostock:** View Stock. **385 Foster + Partners. 386 age fotostock:** jovannig. **388 Corbis:** Noboru Hashimoto / Sygma (bc).

MAPS

Maps are provided throughout the book to illustrate selected railroad lines and other features. Please note that these are for general information only and are not intended to be comprehensive. Place names are given according to the period the map depicts.

CONVERSIONS

Unless specified, figures are approximate and are given to the nearest round number. Where currency conversions are given, the conversion is calculated to the approximate exchange rate of the period, unless specifically stated otherwise.

NUMBERS

"Billion" indicates the short-scale definition, or 1,000 million.